T0033454
RUSSIA
FINLAND
SWEDEN
ESTONIA
LATVIA
LITHUANIA
BELARUS
POLAND
CZECH REP.
SLOVAKIA
UKRAINE
AUSTRIA
HUNGARY
ROMANIA
BULGARIA
GREECE
MALTA
CYPRUS
LEBANON
ISRAEL
SYRIA
TURKEY
GEORGIA
KAZAKHSTAN
MONGOLIA
UZBEKISTAN
KYRGYZSTAN
TURKMENISTAN
TAJIKISTAN
CHINA
NORTH KOREA
SOUTH KOREA
JAPAN
IRAQ
IRAN
AFGHANISTAN
PAKISTAN
NEPAL
BHUTAN
INDIA
KUWAIT
QATAR
UAE
OMAN
SAUDI ARABIA
YEMEN
LIBYA
EGYPT
CHAD
SUDAN
ERITREA
DJIBOUTI
SOMALIA
ETHIOPIA
SOUTH SUDAN
CENTRAL AFRICAN REPUBLIC
CAMEROON
GABON
REPUBLIC OF THE CONGO
CONGO (DEMOCRATIC REPUBLIC)
UGANDA
KENYA
RWANDA
BURUNDI
TANZANIA
ANGOLA
ZAMBIA
MALAWI
MOZAMBIQUE
NAMIBIA
ZIMBABWE
BOTSWANA
SOUTH AFRICA
MADAGASCAR
COMOROS
SEYCHELLES
MAURICIO
REUNION
MALDIVES
SRI LANKA
MYANMAR (BURMA)
VIETNAM
LAOS
THAILAND
CAMBODIA
TAIWAN
PHILIPPINES
MALAYSIA
BRUNEI
SINGAPORE
INDONESIA
EAST TIMOR
PAPUA NEW GUINEA
SOLOMON ISLANDS
VANUATU
NEW CALEDONIA
FIJI
AUSTRALIA
NEW ZEALAND

BRILLIANT MAPS FOR CURIOUS MINDS

100 New Ways to See the World

Originally published in the UK by Granta Books in 2019 under the title *Brilliant Maps*.
First published in North America in revised form by The Experiment, LLC, in 2019.

The Experiment, LLC
220 East 23rd Street, Suite 600
New York, NY 10010-4658
theexperimentpublishing.com

Library of Congress Cataloging-in-Publication Data

Names: Wright, Ian (Blogger), author. | Infographic.ly (Firm), illustrator.
Title: Brilliant maps for curious minds : 100 new ways to see the world /
  Ian Wright ; illustrated by Infographic.ly.
Other titles: Brilliant maps
Description: New York : The Experiment, 2019.
Identifiers: LCCN 2019031051 (print) | LCCN 2019031052 (ebook) | ISBN
  9781615196258 | ISBN 9781615196265 (ebook)
Subjects: LCSH: Human geography--Maps. | Atlases. | Curiosities and
  wonders.
Classification: LCC G1046.E1 W75 2019 (print) | LCC G1046.E1 (ebook) |
  DDC 912--dc23
LC record available at https://lccn.loc.gov/2019031051
LC ebook record available at https://lccn.loc.gov/2019031052

ISBN 978-1-61519-625-8
Ebook ISBN 978-1-61519-626-5

Cover and text design by Beth Bugler
Maps by Infographic.ly

Manufactured in Turkey

First printing November 2019
10 9 8

BRILLIANT MAPS FOR CURIOUS MINDS

100 New Ways to See the World

IAN WRIGHT
Illustrated by Infographic.ly

CONTENTS

PEOPLE AND POPULATIONS

POLITICS, POWER, AND RELIGION

CULTURE AND CUSTOMS

FRIENDS AND ENEMIES

GEOGRAPHY

HISTORY

NATIONAL IDENTITY

CRIME AND PUNISHMENT

9

NATURE

INTRODUCTION

If you love maps, trivia, or just learning more about our world, then I think you'll love this book. It takes its lead from my popular Brilliant Maps website (brilliantmaps.com), which curates, collects, and contextualizes maps from around the internet. Since launching in late 2014, I've published nearly 350 maps, which have been viewed by nearly 15 million people; that's more than the population of Somalia, Cuba, Belgium, or Greece.

Perhaps even more amazing is the truly global reach of the maps. The website has received visitors from 241 countries and territories, which is impressive considering the United Nations only has 193 member states. People have visited the site from such far-flung places as Antarctica, North Korea, Western Sahara, Christmas Island, and the British Indian Ocean Territory (an archipelago halfway between Tanzania and Indonesia with no permanent residents). People have logged on from 34,262 towns and cities worldwide including Washington (England), Ottawa (Illinois), and Sydney (Nova Scotia). All these statistics show that curiosity about the world around us is something that we all have in common.

On my blog, I've had a behind-the-scenes window into just how popular certain maps and topics are. The book is largely a collection of the most talked-about maps, the ones that seemed especially to strike a nerve with readers.

Unsurprisingly, in the last five years, the maps that get the most comments are almost all related to national identity or politics. And let's just say that people have very strong opinions on these subjects. When I started the blog, it was a simpler time; Barack Obama was still in the White House, and few took either Brexit or Donald Trump seriously. Now here we are. Of the ten most viewed maps on the website, two are related directly to the 2016 US presidential election, and a further three deal with issues of European national identity and immigration.

I think it's worth looking at a few of the popular and controversial maps from the website. The most popular map on the website both by visits and comments is "Second-largest nationality living in each European country" (see page 10) and shows exactly what it says on the package. At first glance, there's no reason why this map should be particularly popular besides being rather nice looking (it uses flags instead of country names to present the data). However, I think its popularity can be attributed to a few factors. The first is genuine surprise (for example, who knew the second-largest nationality in Portugal were Brazilians, rather than any of the neighboring European countries?). But another reason is that people disagree with what should be the second-largest nationality. Sometimes there is a disagreement based on the data source (our version uses data from the UN Department of Economic and Social Affairs), while in other cases it doesn't fit with people's mental model of how they think the world or their country works.

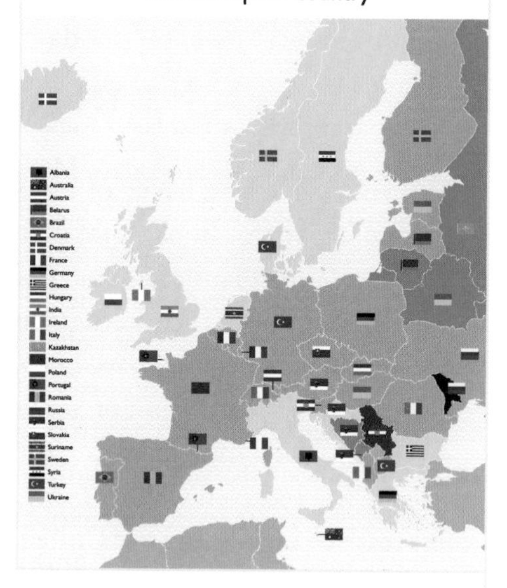

The third and most sinister reason the map proved popular is the fact that it shows nationals from several predominately Muslim countries as the second-largest nationality in a number of non-Muslim countries (e.g., Algerians in France, Turks in Germany and several other countries, Moroccans in Belgium). This plays on right-wing fears that Europe is somehow being "overrun" with Muslims, when in actual fact Muslims only make up roughly 6 percent of Europeans and 4 percent of the European Union. These fears also mean that the post gets shared widely on far-right Twitter and Facebook groups (not something I'm in any way happy about but can't really prevent).

Another map has become popular not because it's widely shared on social media but because it covers topics people frequently search for: "England vs.

Great Britain vs. United Kingdom" (see page 29), which simply shows the difference between the United Kingdom, Great Britain, Ireland, and England. Over five hundred people per day visit the site to try to understand these differences.

To paraphrase, there are "Lies, damned lies, and maps." Or to put it another way, maps can easily mislead. Most of the maps I share on the site are generally comparing countries based on a single dimension. This means some context is often not only helpful but necessary to understand the map. If someone is convinced that Trump won the popular vote, seeing a map showing the result by county will confirm that belief, fake news or not. Similarly, any sort of map dealing with genetics is almost certain to attract racists who will claim the superiority of their own group over others. People see what they want to see. This of course creates a dilemma on my part. On the one hand, I could choose not to share these maps and avoid any sort of controversy. On the other hand, these maps are being created and shared anyway; by providing context I can hopefully help shape the debate.

Ultimately, what I've opted to do is include some of these maps but mix them up with other kinds of maps. I sometimes post humorous maps that explore these divisions and differences in a much more lighthearted way. After publishing 350 maps, I have a fairly good idea of what will end up being popular. Anything with a political or national identity angle has a vastly greater chance of going viral than a map about the average color of each country's flag, because we have a boundless appetite for learning more about ourselves and our world.

It's very difficult to pick personal favorites since every map has something about it I find interesting. Nevertheless, there are two types of maps that I really enjoy: ones that are surprising and/or ones that tell an interesting story. Among these are population maps, like the ones I've included for Canada and Australia (see pages 6–7). When looking at either country on a map it's clear that they have a lot of geographical space; they are the second and sixth largest countries in the world by area, respectively. But, as both of these maps show, most people cluster in very small sections of each country. Most of that empty space is uninhabited or very sparsely populated. If you ever visited the center of Australia or the northern reaches of Canada, you would quickly understand why. Similarly, "Countries with economies bigger than California's" (see page 36) really strikes me. If California separated from the United States, it would not only be the thirty-fifth most populous country on Earth but also have the world's seventh largest economy. Finally, there are two physical geography maps included that reveal some unexpected truths. "Chile is a ridiculously long country" (see page 110) really changes our perspective on Chile's size; held up against European landmasses, it becomes clear just how long it really is, and indeed how huge South America must be to accommodate it. "Countries with no rivers" (see page 176) also promises a shift in the way we look at resources and geography: While you might expect a few small countries not to have rivers, there are actually some really big ones that do not contain a single river.

Chile is a ridiculously long country

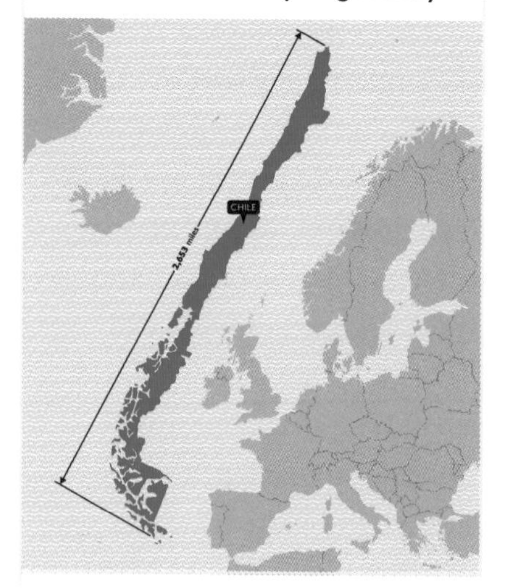

In terms of interesting stories, there are several I've included that I particularly like. The first is "Travel times from London in 1914," which pairs with "Travel times from London in 2016" (see pages 126–29), isochron maps showing how long it takes to get to any other point in the world from London. It's an amazing visualization of how air travel has changed our ability to reach other parts of the world and expand our horizons. You can now get to almost any other point in the world from London in around thirty-six hours, whereas a century ago it could have easily taken thirty-six days to get to some of the remote locations. The second is a pair of maps that tell a very thought-provoking story: "Where North Korea has embassies" and "Who has embassies in North Korea?" The assortment of countries range from the expected (Communist China, North Korea's main ally, and Russia) to the unexpected (such as the UK and Sweden). However, if you study the maps closely, you'll see that not all countries that host a North Korean embassy have one in Pyongyang.

Map of the **entire internet** in December 1969

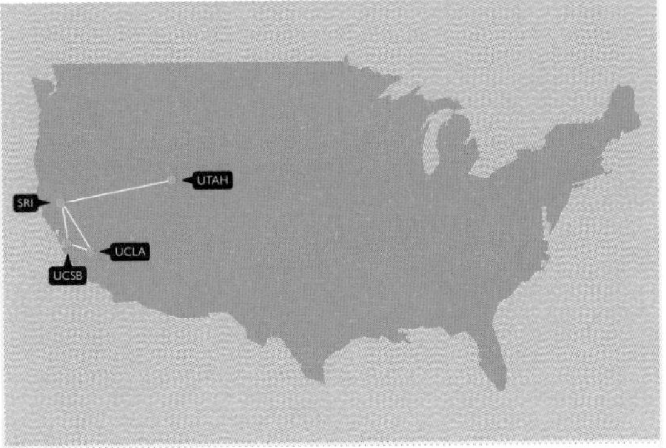

The "Map of the entire internet in December 1969" is one of my all-time favorites, especially since this book would not exist without it. What's interesting is not so much how quickly the internet has grown but how long it took to happen. It wasn't until 1989 that the web (one of the ways for people to access the internet) was born, two decades after this map. However, in the following twenty years, all of the internet giants were founded, from Amazon to Facebook to Google. It is a good reminder that truly revolutionary change can take a long time to get started but then take off very quickly once it does.

My all-time favorite map from the website (not included here only for rights reasons) is the "1988 East German map of West Berlin." Besides being a German history buff, the reason why I think this map is so amazing is that it simply shows a big white blob where West Berlin should be. At first glance, the map is ridiculous and useless and obviously wouldn't fool anyone. However, upon further reflection, it's not as silly as it seems, because for most East German citizens, showing West Berlin on a map would have been as useful as showing the moon. They couldn't travel there, so why show it? What the map actually shows us is where East German citizens were able to travel. There's also a similar map showing the East Berlin U-Bahn and S-Bahn network that completely omits West Berlin (although this is not shown as a big white blob). In my opinion, this 1988 map is near perfect. It is an accurate map, showing historical fact, while also being completely absurd at the same time.

I do hope you enjoy this book, and if you haven't checked out the Brilliant Maps website yet, I hope to see you there for even more brilliant maps.

IAN WRIGHT
London, 2019

1

PEOPLE AND POPULATIONS

1 European countries overlaid on areas of the Americas with **equal populations**

2 US states* overlaid on areas of Europe with **equal populations**

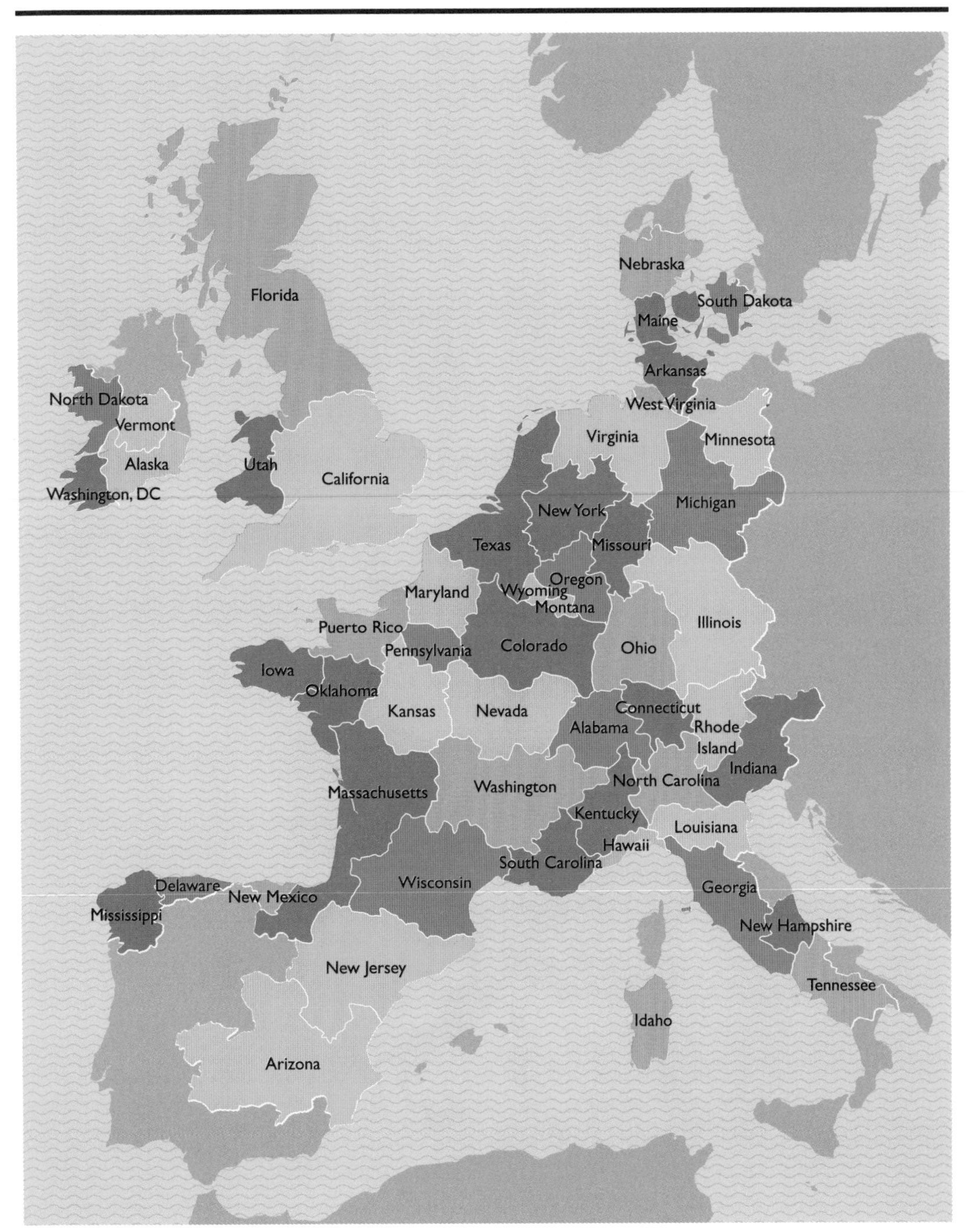

* Plus Washington, DC, and Puerto Rico

3 How the North American population **fits into Europe**

US population: 326,204,292

Mexican population: 130,558,054

Canadian population: 36,912,135

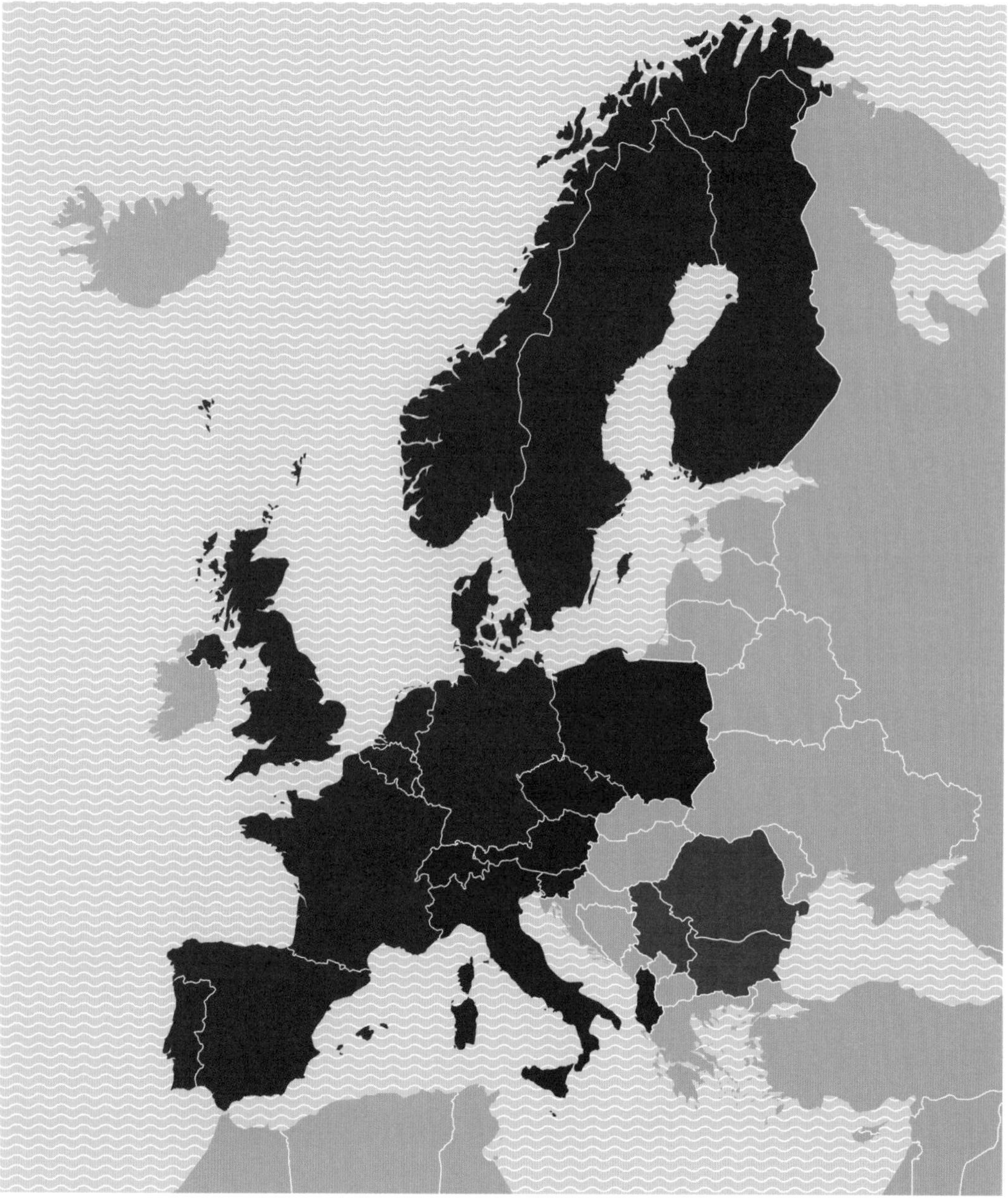

4 The astounding drop in global **fertility rates** from 1970 to 2015

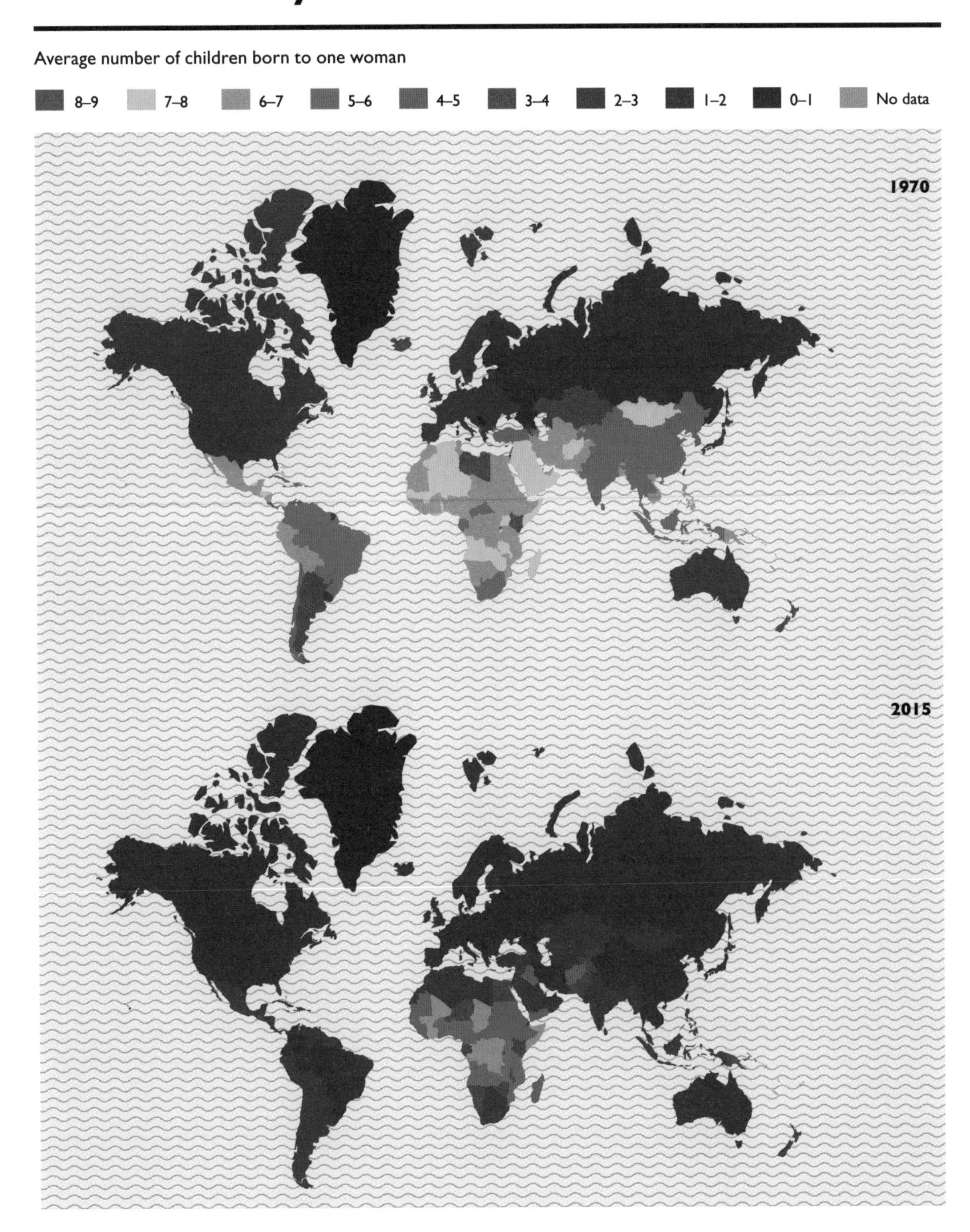

5 More than half of the **Australian population** lives here

6 50% of Canadians live south of the red line

While the forty-ninth parallel is often thought of as the border between the US and Canada, the vast majority of Canadians (roughly 68 percent) live below it, with more than 50 percent of Canadians living south of 45°42' (the red line).

Toronto (43°42'N) and Montreal (45°30'N), Canada's biggest and second-biggest cities, respectively, both lie just below the line, as does Ottawa (45°25'N), the capital and the fourth-largest city.

7 Countries with the largest **immigrant populations**

<1%

1–3.9%

4–9.9%

10–19.9%

20–49.9%

>50%

N/A

8 **Second-largest nationality** living in each European country

No data | 5–9% | 10–19% | 20–29% | 30–39% | 40–49%

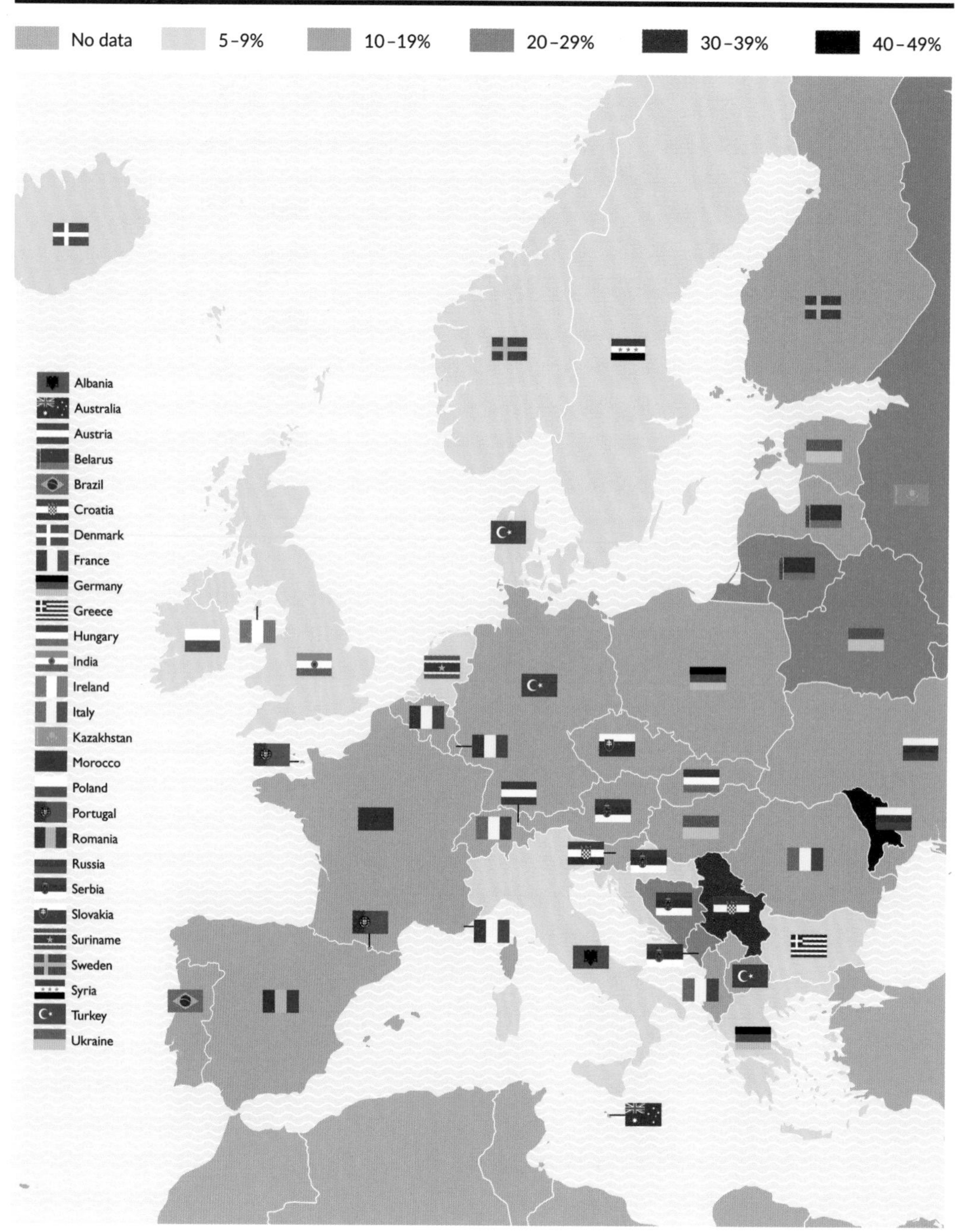

* In 2017, including Russia

9 Percentage of people born in each European country now **living abroad**

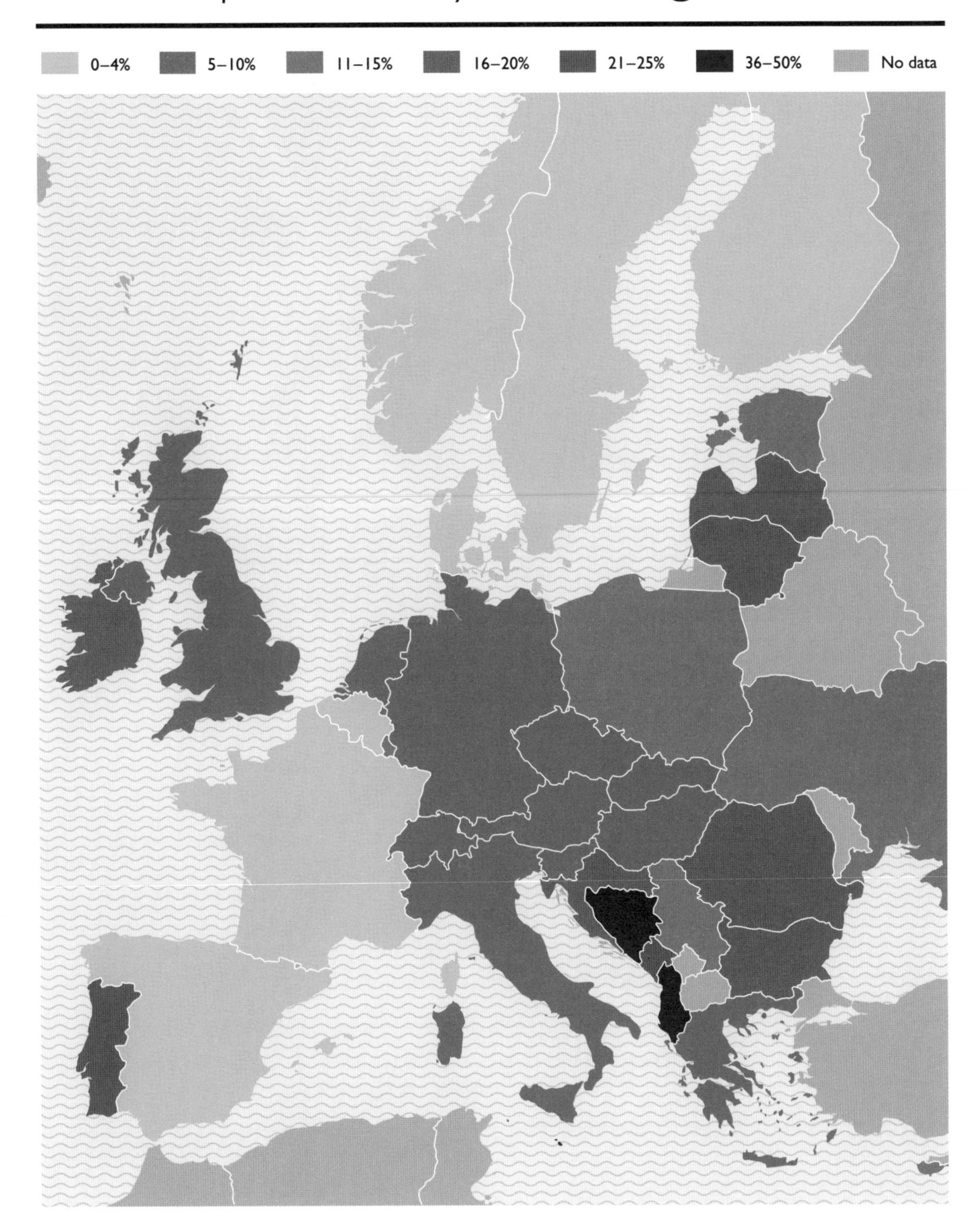

10 Countries and overseas territories that each have smaller populations than **Greater Tokyo***

* 37.8 million

TOKYO

11 World **median ages**

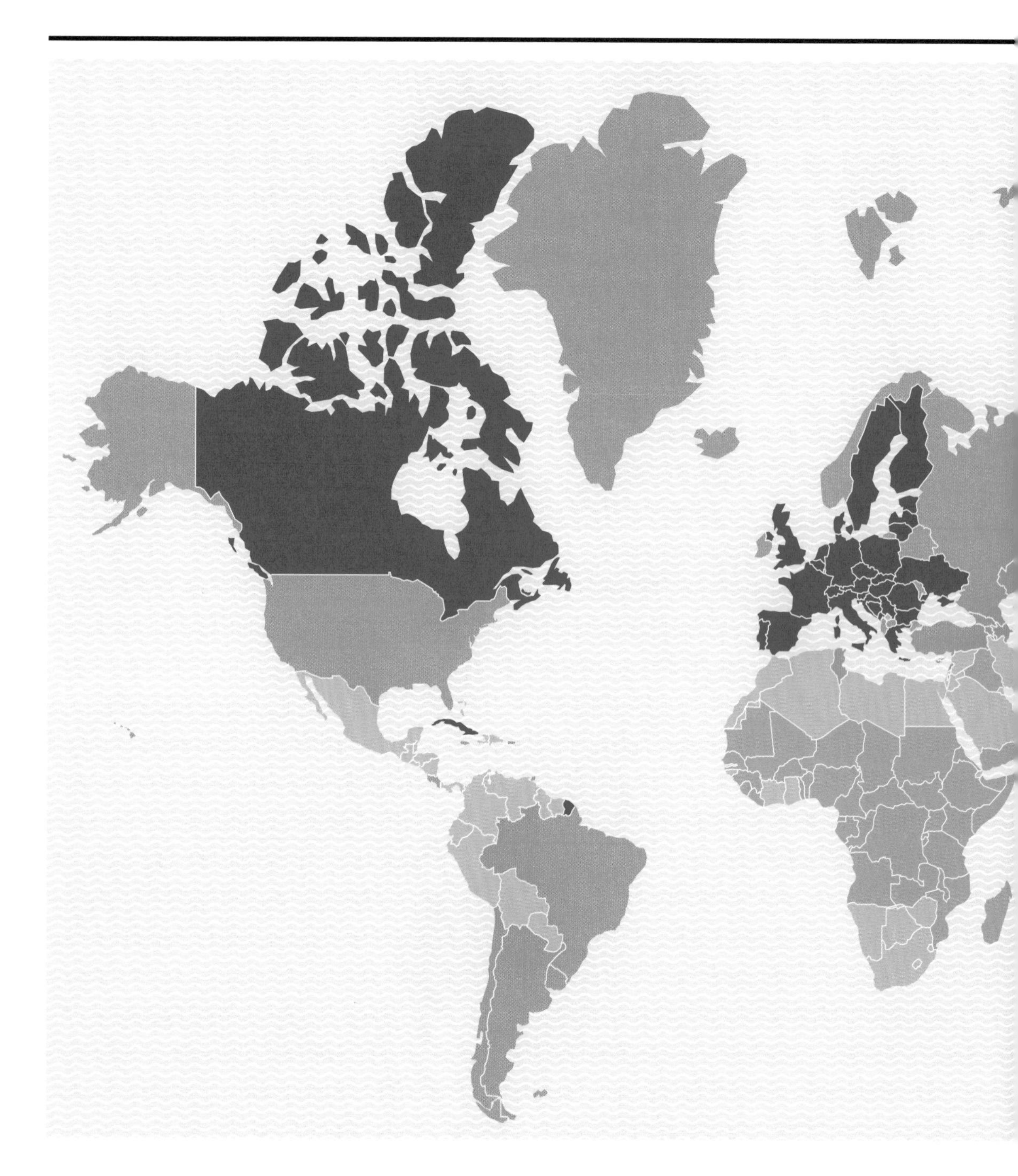

Median age

- Teens
- Twenties
- Thirties
- Forties

Youngest

1. Niger **15.3**
2. Uganda **15.7**
3. Mali **16.2**
4. Malawi **16.5**
5. Zambia **16.7**

Oldest

1. Japan **46.9**
2. Germany **46.8**
3. Italy **45.1**
4. Greece **44.2**
5. Slovenia **44.1**

12 Average **female height** worldwide

5 ft. 6 in.
5 ft. 4 in.
5 ft. 2 in.
5 ft.
4 ft. 10 in.
No data

13 Average **male height** worldwide

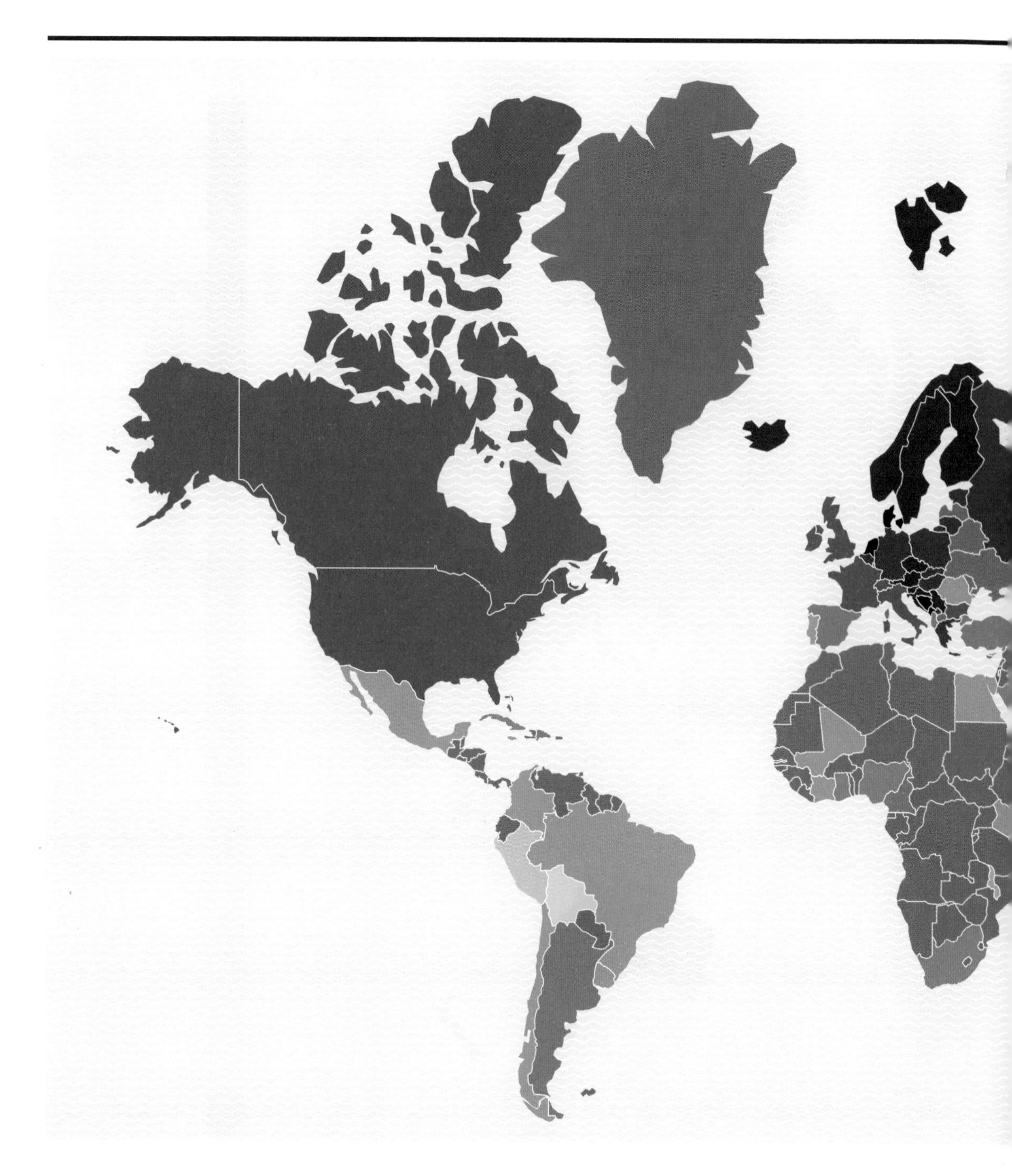

6 ft.
5 ft. 10 in.
5 ft. 8 in.
5 ft. 6 in.
5 ft. 4 in.
5 ft. 2 in.
No data

2

POLITICS, POWER, AND RELIGION

14 64 countries have had a **female leader of government** in the last 50 years

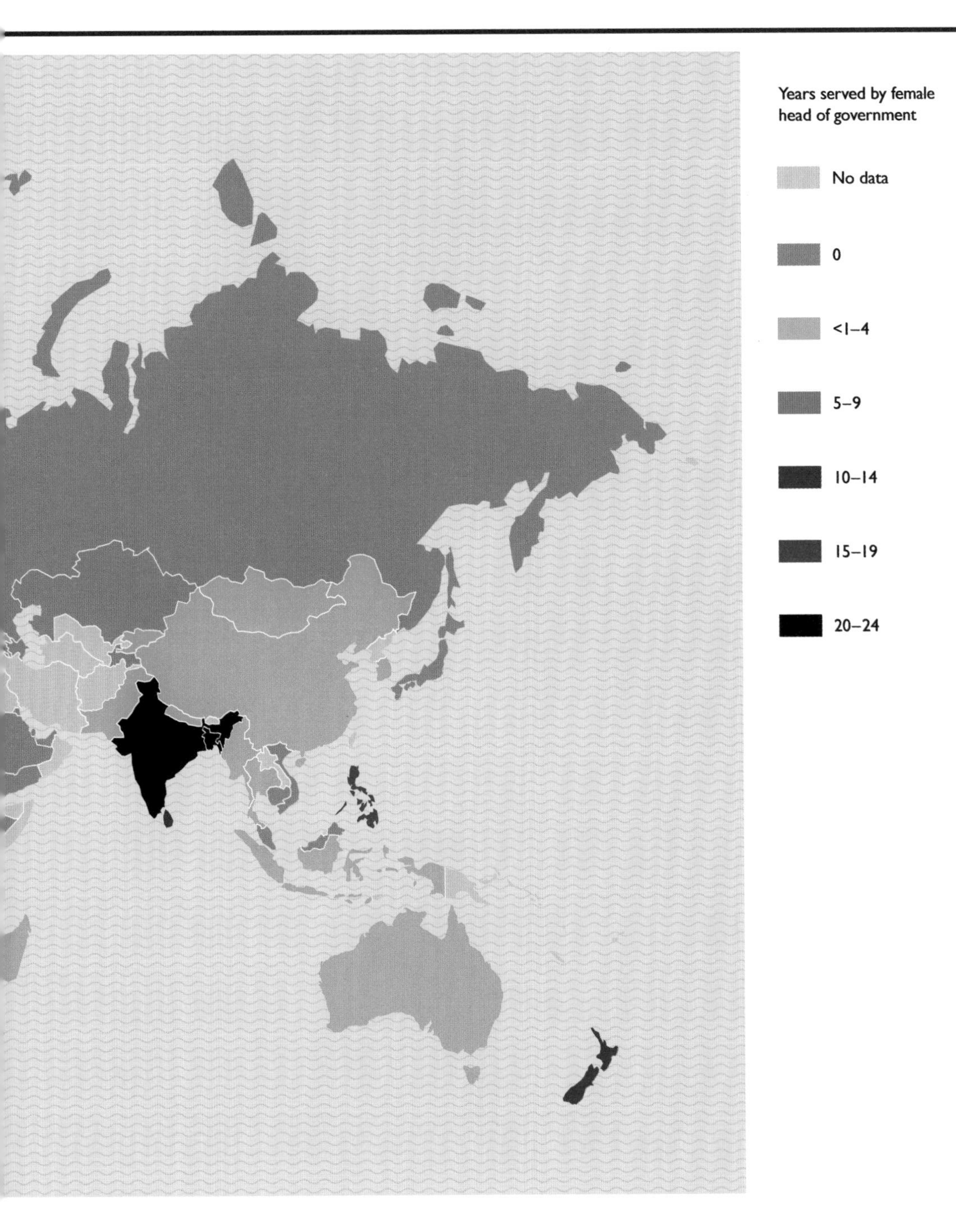
Years served by female head of government
No data
0
<1–4
5–9
10–14
15–19
20–24

15 Countries with **state religions**

Israel is defined as a "Jewish and democratic state," but as the term "Jewish" describes the Jewish people as both an ethnic and religious group, it is not included on this map.

16 Fastest-growing religion

in each country around the world

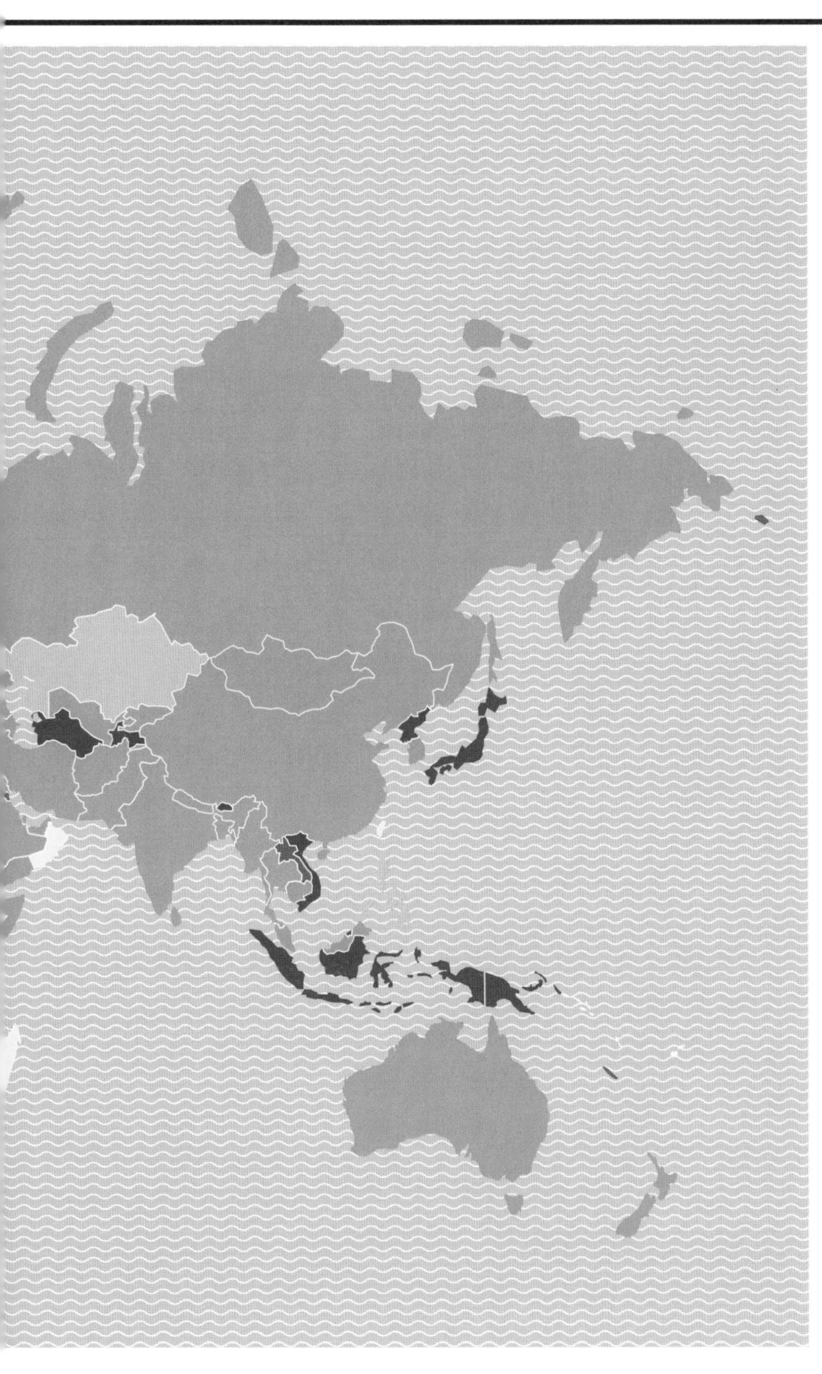

Buddhism

Christianity

Folk
(religion outside official doctrine or practice)

Hinduism

Judaism

Islam

Other

Unaffiliated

17 Simplified map of **Africa's religions**

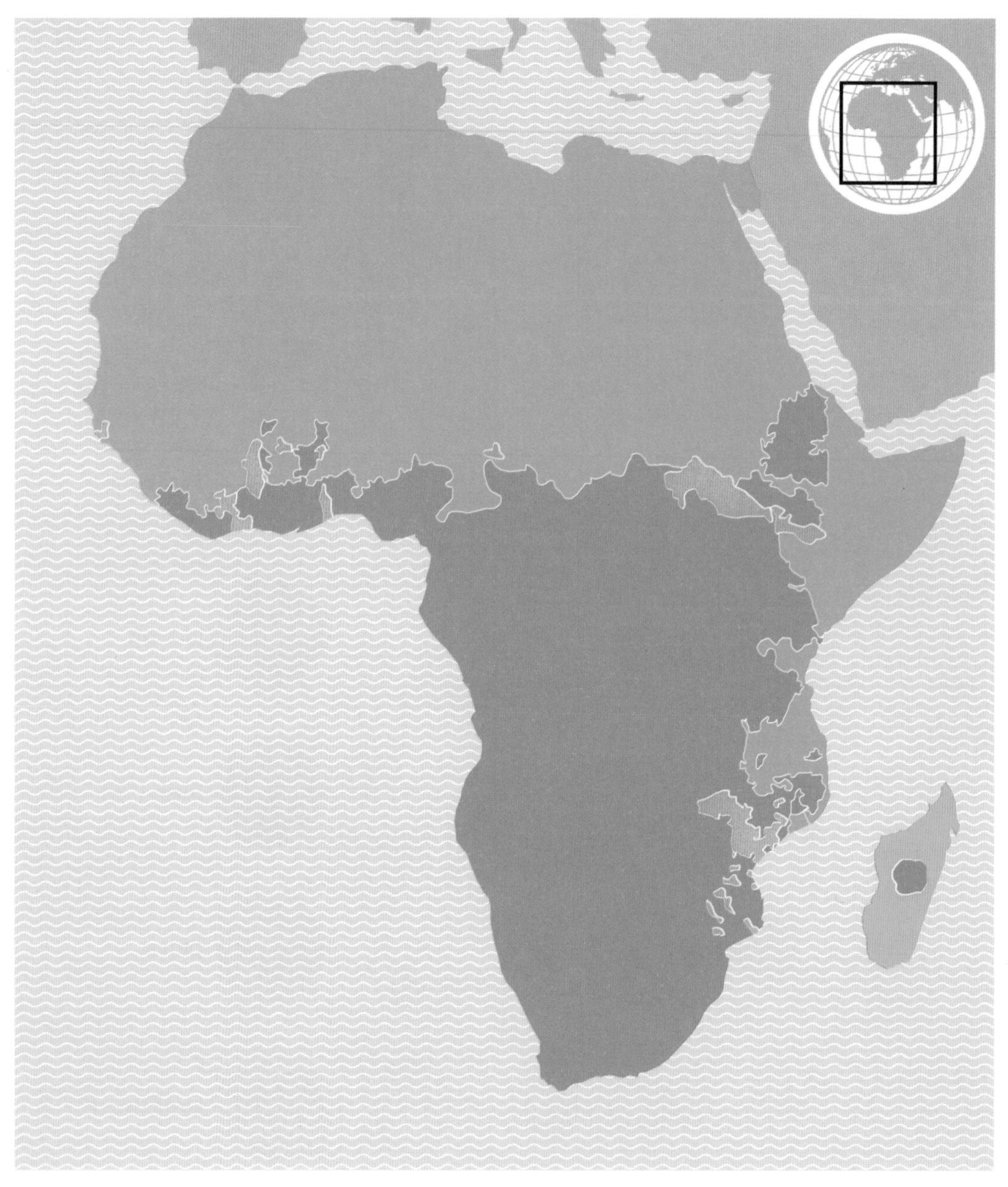

18 England vs. Great Britain vs. United Kingdom

British-Irish Isles
Ireland
United Kingdom
Great Britian

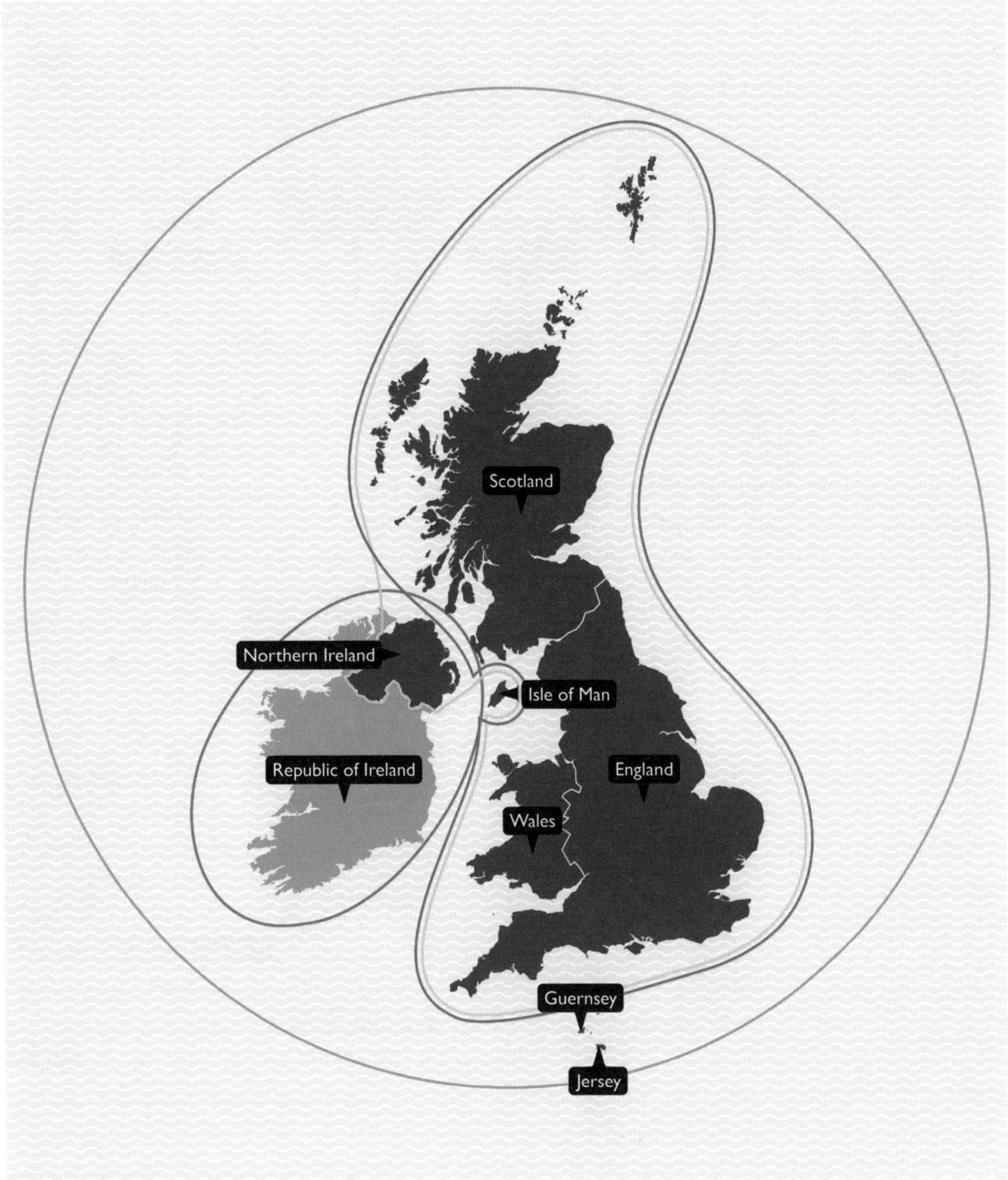

Scotland
Northern Ireland
Isle of Man
Republic of Ireland
England
Wales
Guernsey
Jersey

19 Birthplaces of **religious leaders**

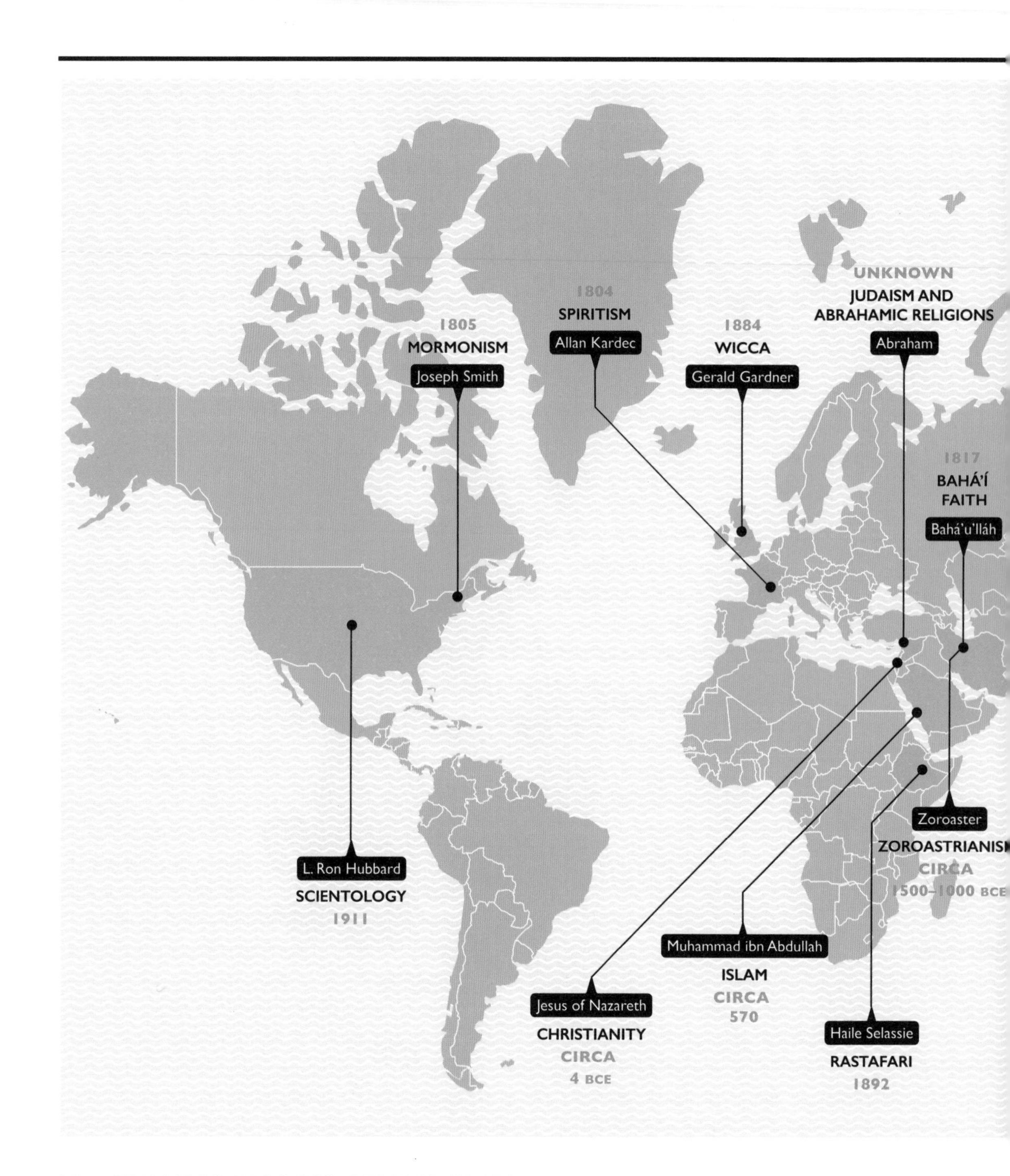

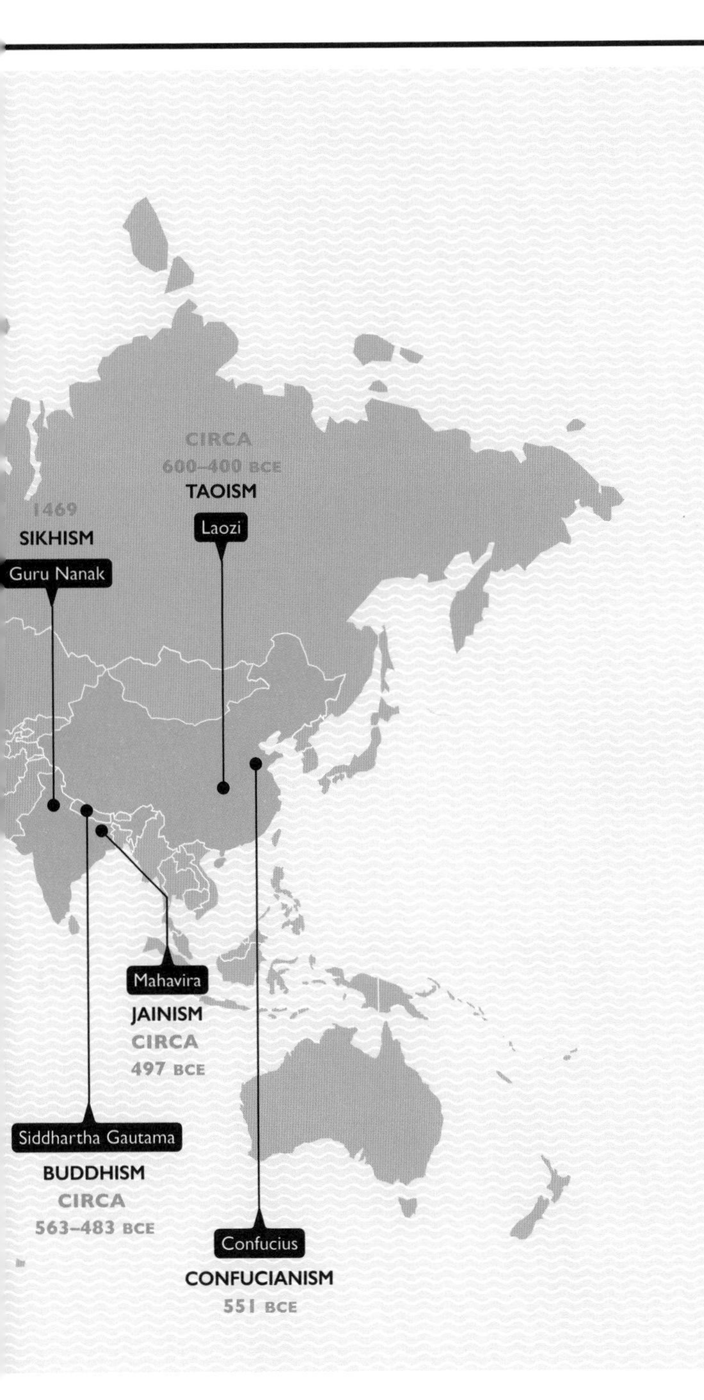

The Saptarishi (Seven Great Yogis) who are considered the founders of the Hindu faith do not have an associated birthplace, though in some parts of India it is believed they are the seven stars of the Big Dipper.

20 Every country's **highest-valued export**

Food/Drink
Metals/Minerals
Precious Metals/
Minerals
Wood Products
Oil and Petroleum
Textiles/Clothing
Machinery/
Transportation
Electronics
Other

21 The largest **source of imports** by country

22 Countries with **economies bigger than California's**

Excluding the US, here in orange are the countries whose economies are bigger than California's (in nominal GDP).

23 World **gold reserves** in grams per person

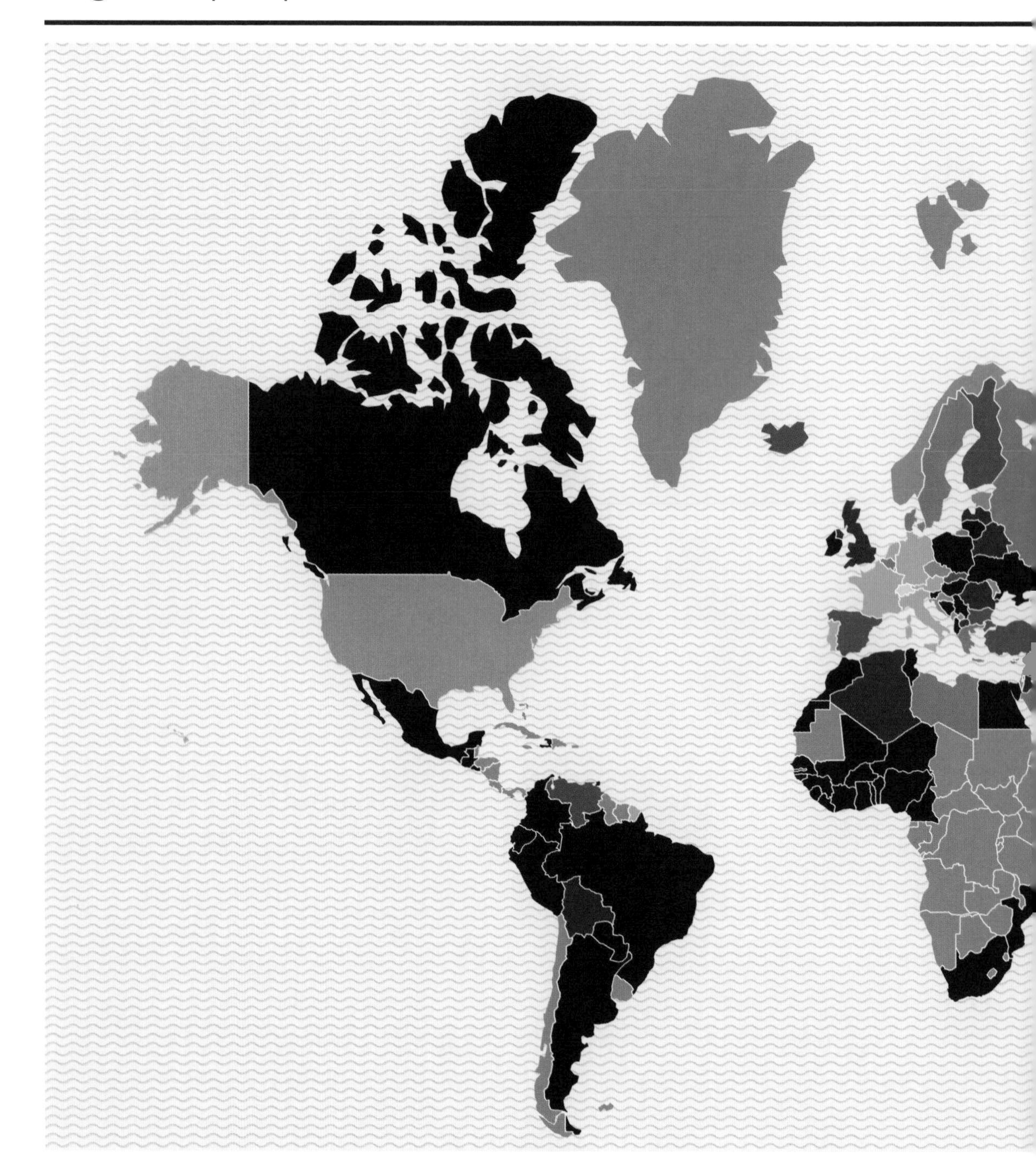

No data
0–3
3–5
5–10
10–20
20–50
50–130

24 World divided in half based on **military spending**

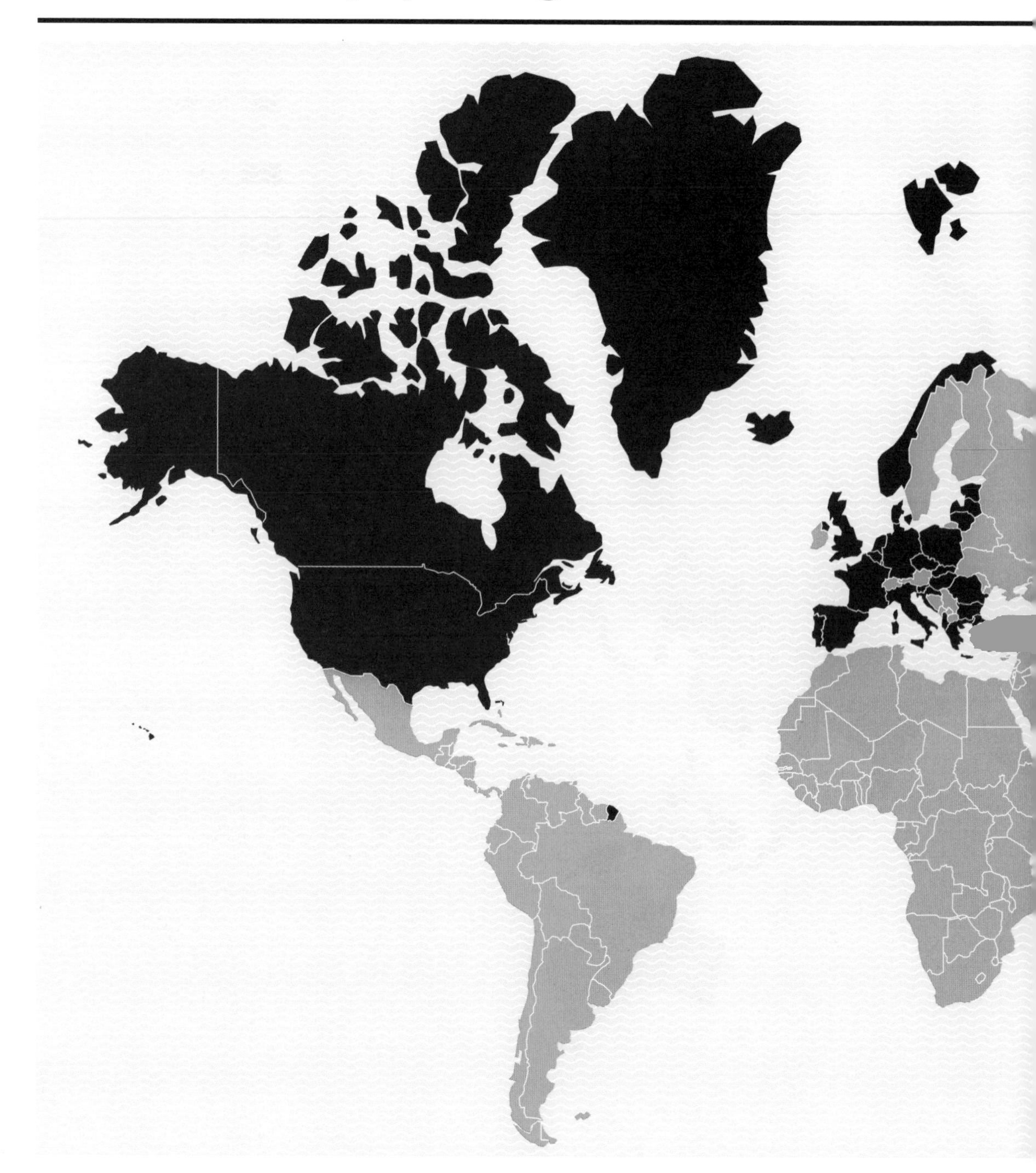

NATO: 54%

The rest: 46%

25 **Nuclear** vs. **non-nuclear** countries*

* In 2018

Generates nuclear power

Does not generate nuclear power

3

CULTURE AND CUSTOMS

26 Who are the world's speed demons? The **highest speed limits** around the world

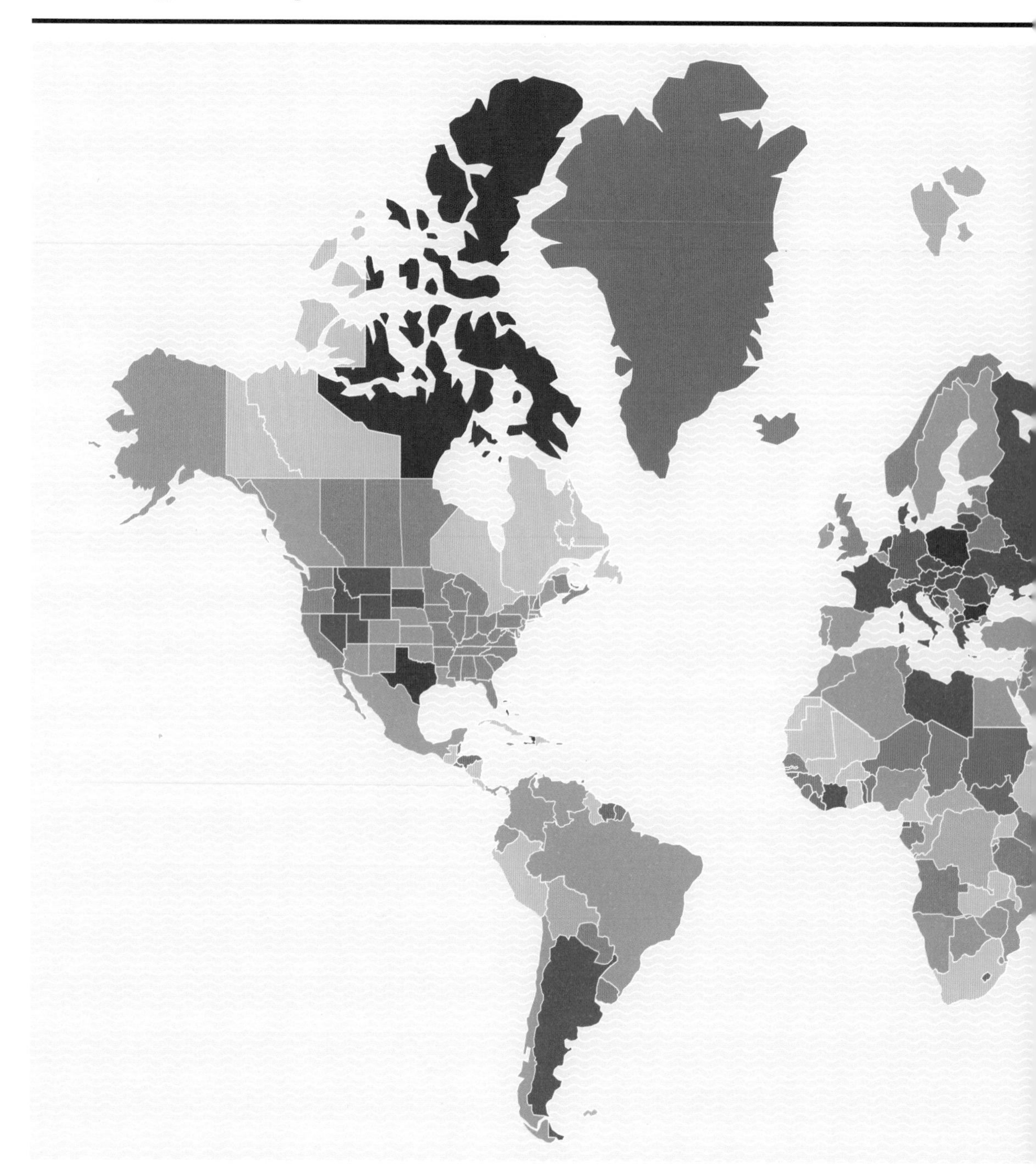

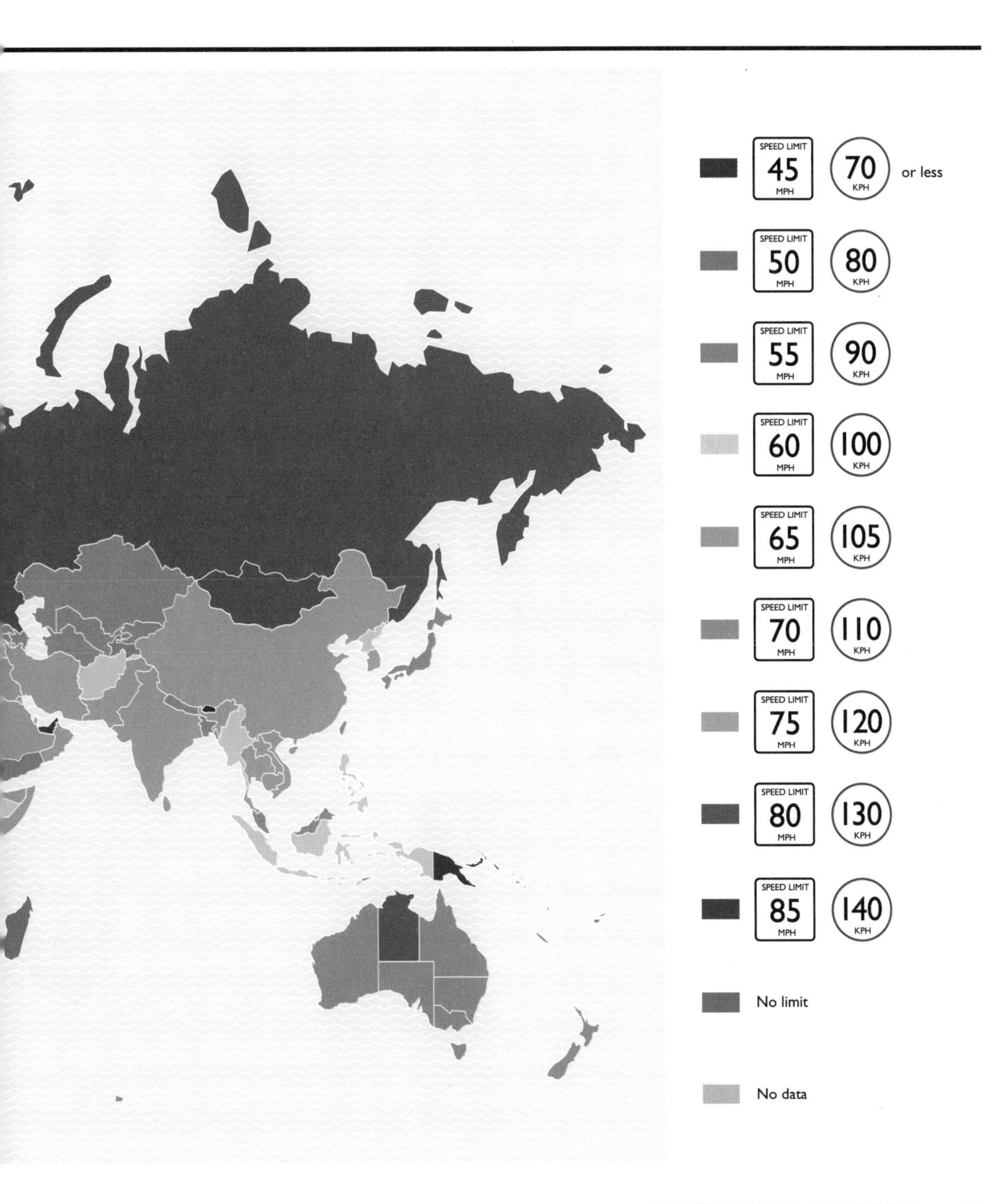
SPEED LIMIT
45
MPH
70
KPH
or less
SPEED LIMIT
50
MPH
80
KPH
SPEED LIMIT
55
MPH
90
KPH
SPEED LIMIT
60
MPH
100
KPH
SPEED LIMIT
65
MPH
105
KPH
SPEED LIMIT
70
MPH
110
KPH
SPEED LIMIT
75
MPH
120
KPH
SPEED LIMIT
80
MPH
130
KPH
SPEED LIMIT
85
MPH
140
KPH
No limit
No data

27 Who drives on the **"wrong" side** of the road?

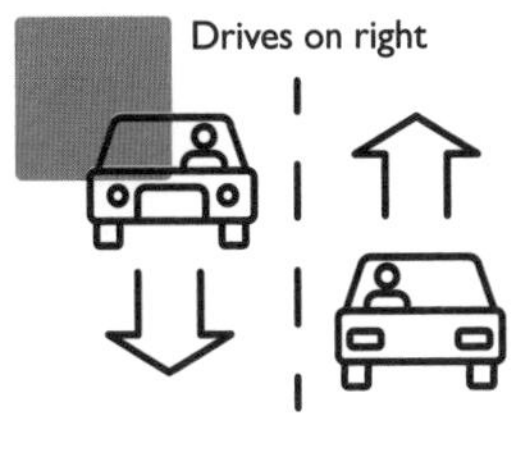
Drives on right

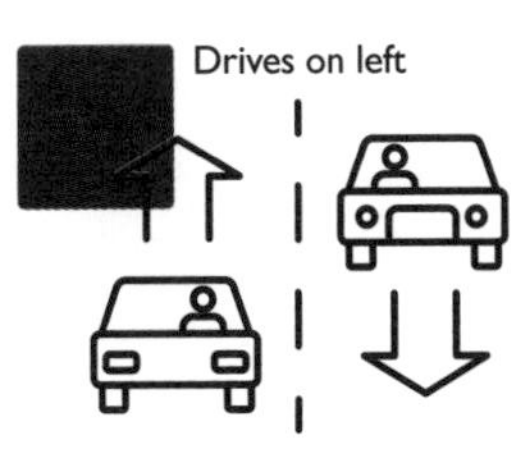
Drives on left

28 Football vs. soccer

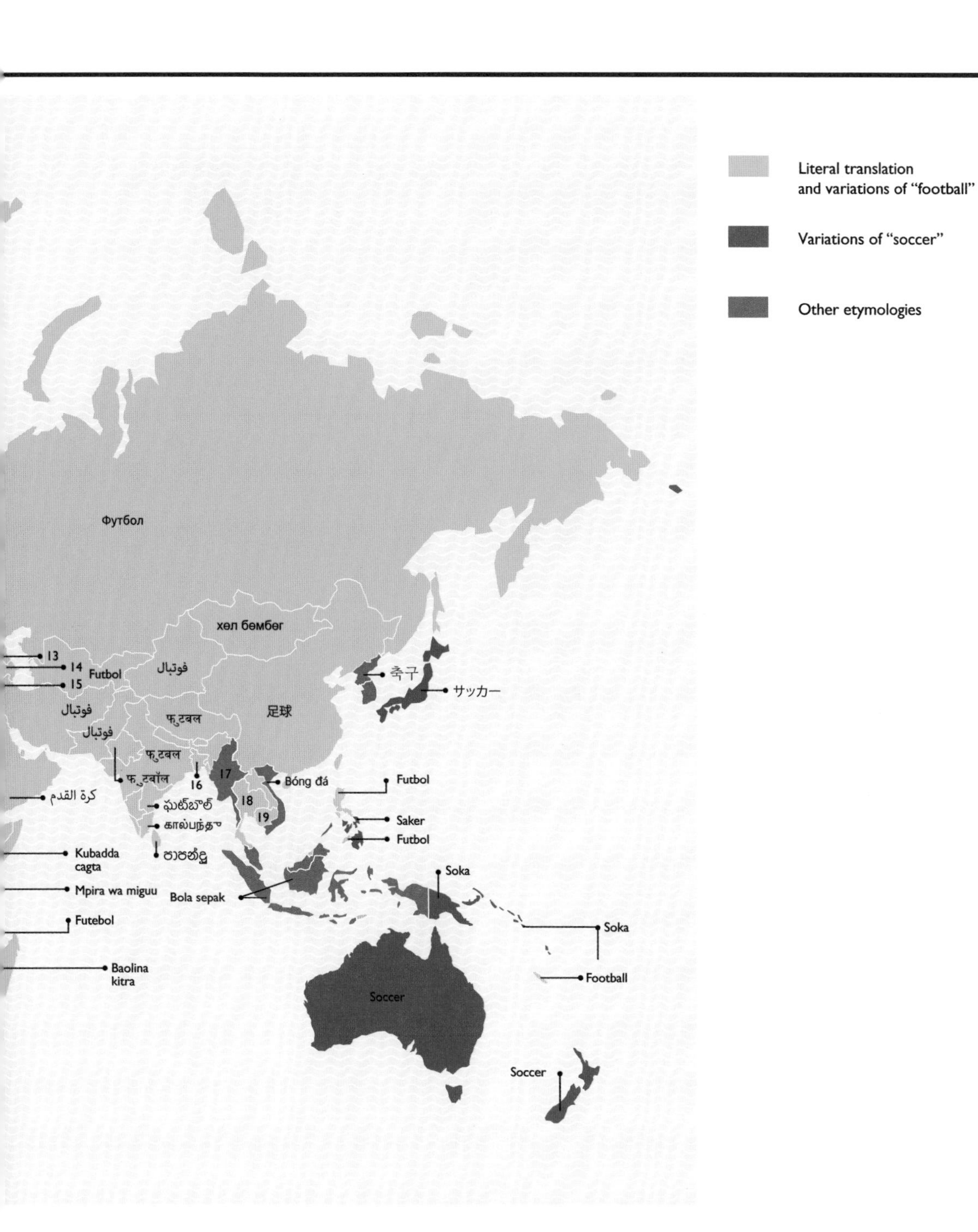

Literal translation and variations of "football"
Variations of "soccer"
Other etymologies
Футбол
хөл бөмбөг
13
14
15
Futbol
فوتبال
فوتبال
فوتبال
फुटबल
足球
축구
サッカー
फुटबल
फुटबॉल
16
17
18
19
Bóng đá
كرة القدم
ఫుట్‌బాల్
கால்பந்து
පාපන්දු
Futbol
Saker
Futbol
Kubadda cagta
Mpira wa miguu
Bola sepak
Soka
Futebol
Soka
Baolina kitra
Football
Soccer
Soccer

29 Most commonly spoken languages
after English and Spanish

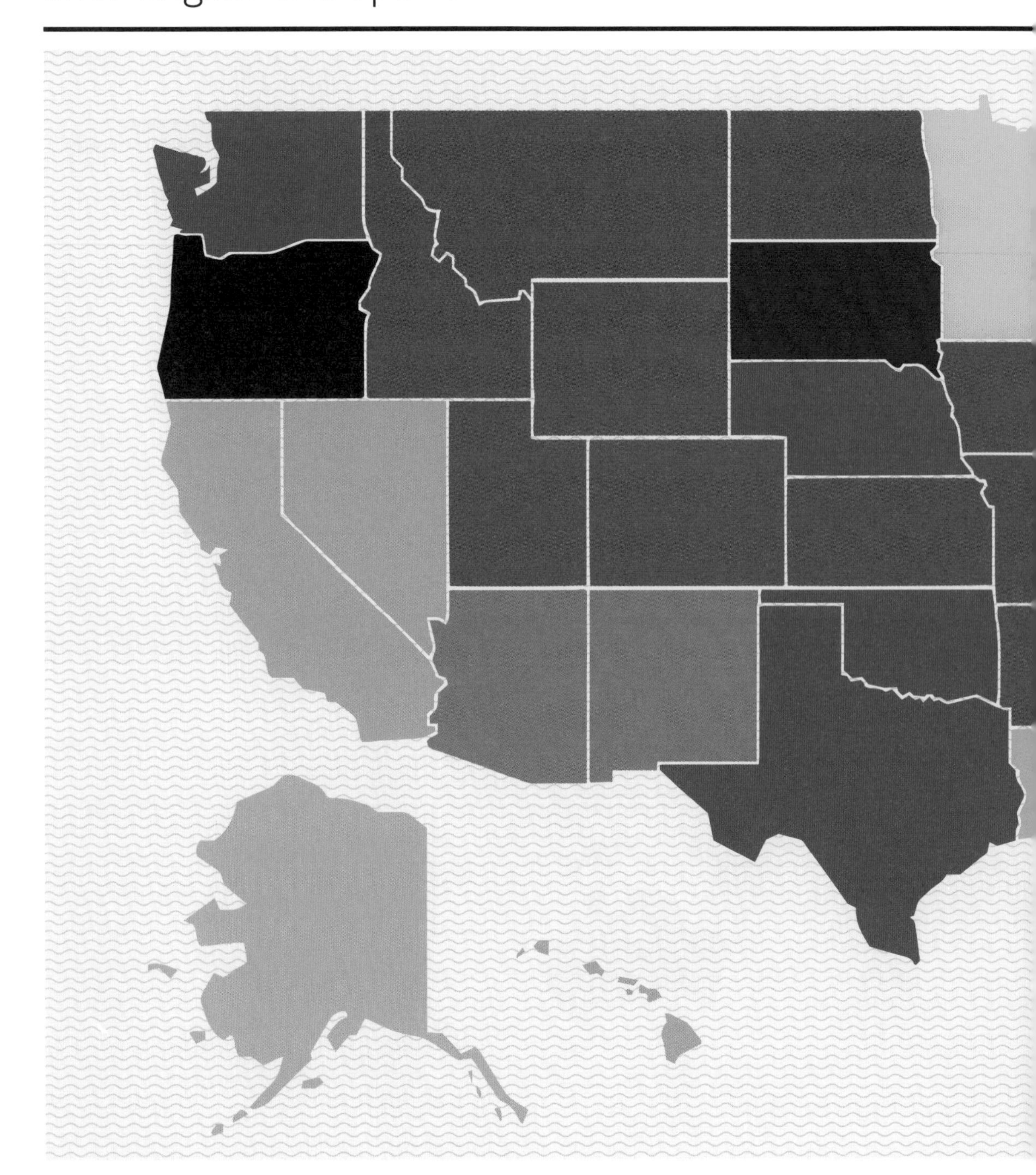

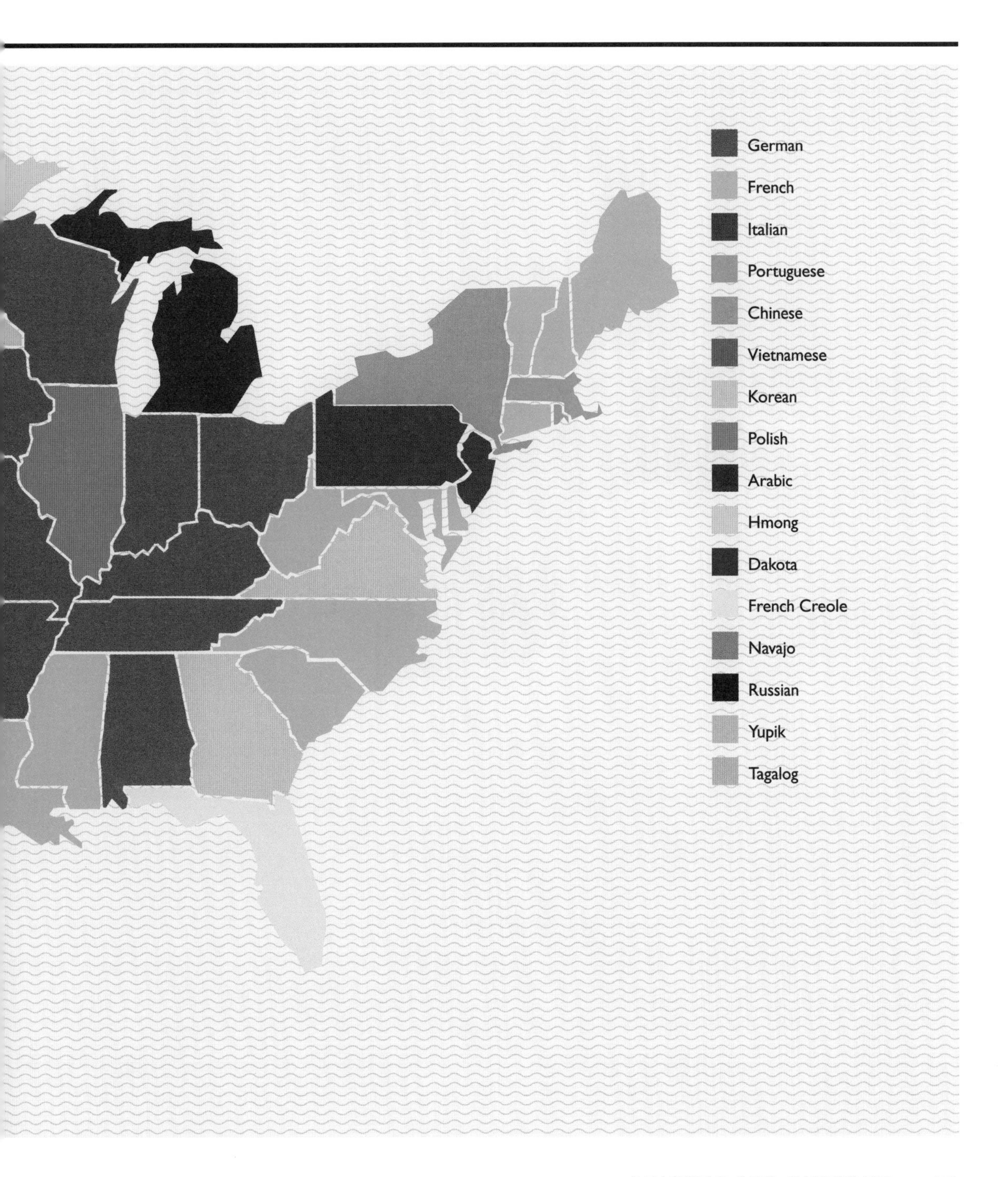
German
French
Italian
Portuguese
Chinese
Vietnamese
Korean
Polish
Arabic
Hmong
Dakota
French Creole
Navajo
Russian
Yupik
Tagalog

30 Map of countries officially **not using the metric system**

31 Decimal point vs. decimal comma vs. other **decimal separators**

Period
0.005

Comma
0,005

Both and/or apostrophe
0.005
0,005
0'005

Momayyez
0,005

No data

32 How to **write the date** in different countries

Year/Month/Day

Day/Month/Year

Month/Day/Year

33 **Election days** by country

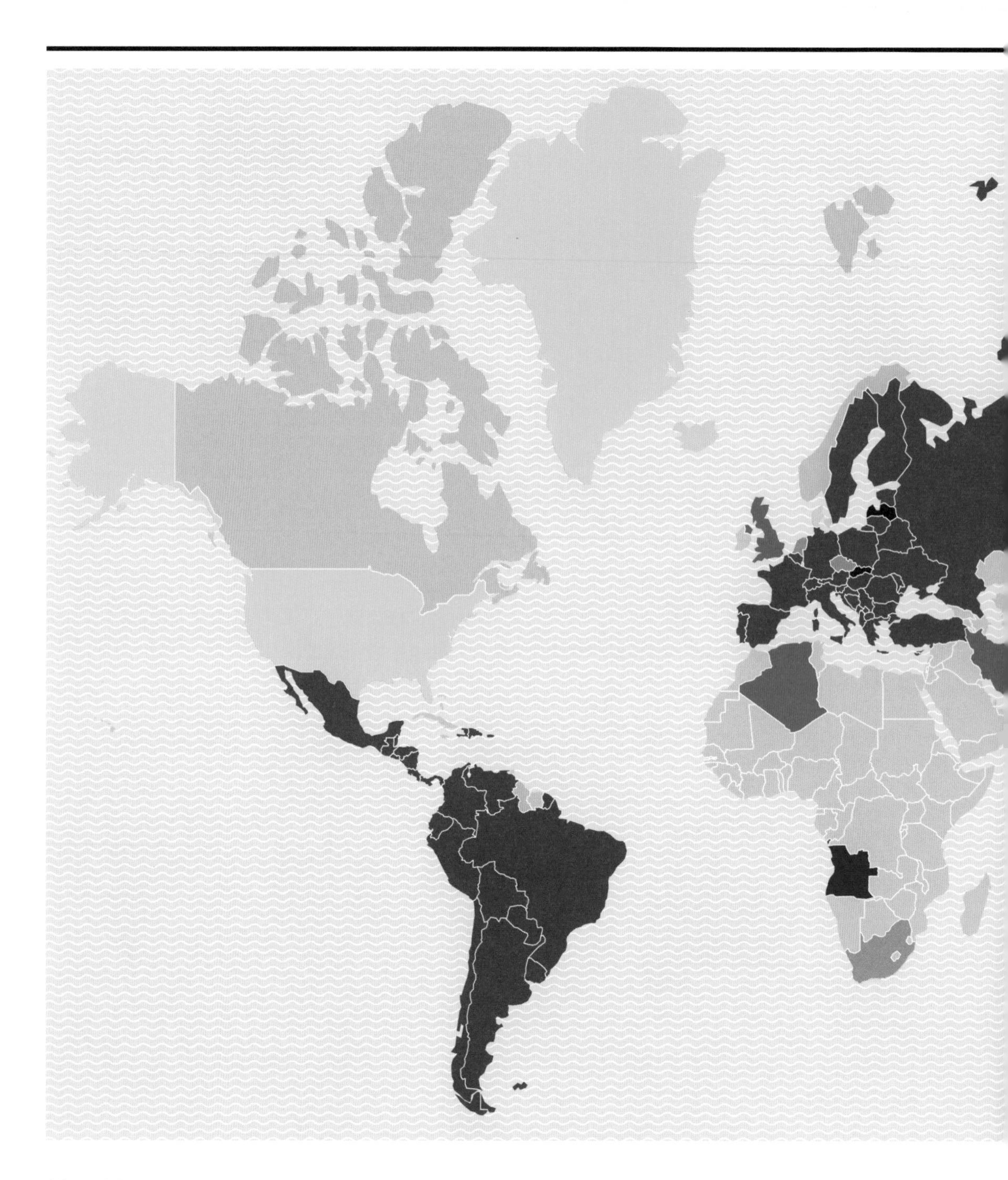

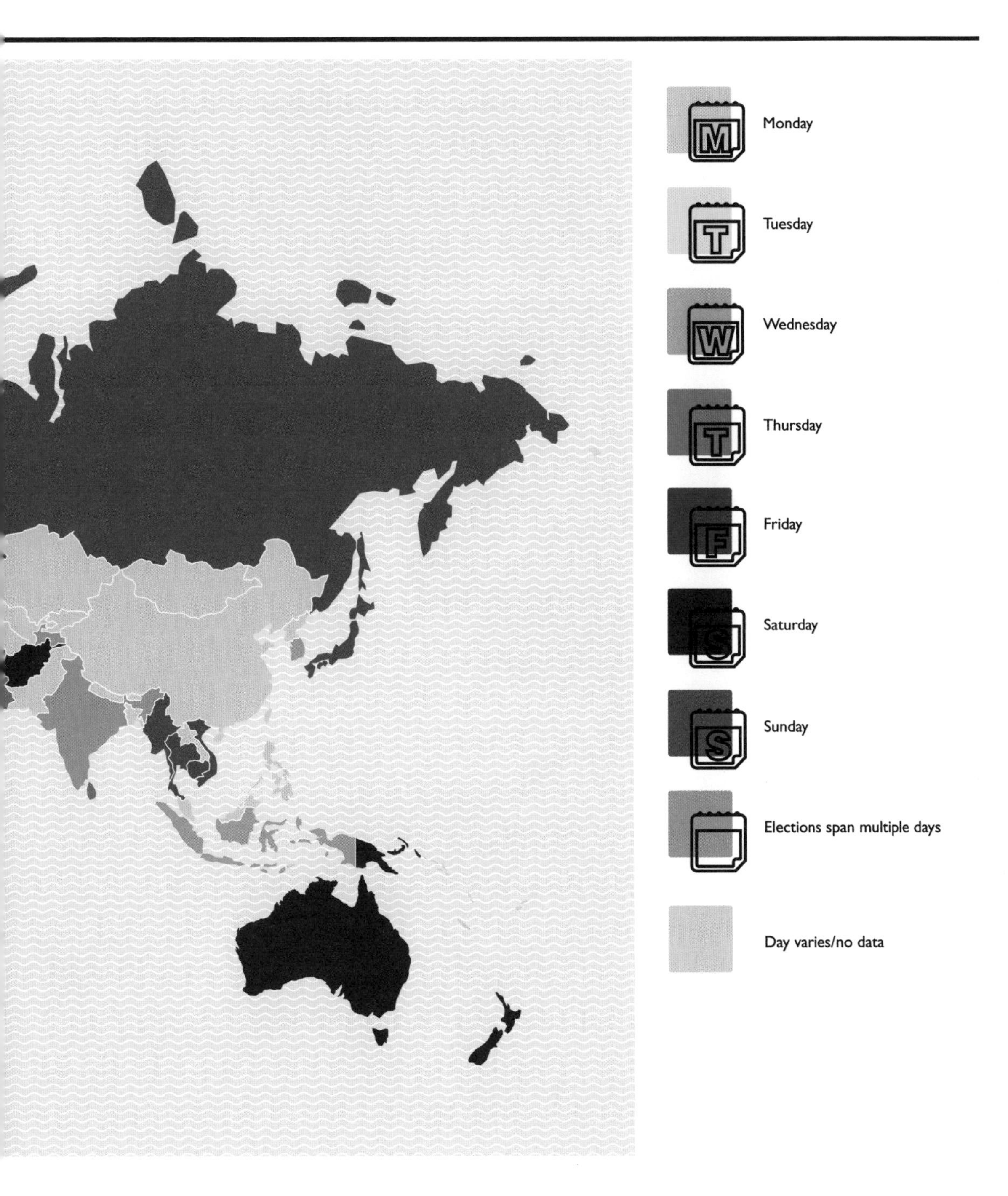
Monday
Tuesday
Wednesday
Thursday
Friday
Saturday
Sunday
Elections span multiple days
Day varies/no data

34 Countries that have **no McDonald's**

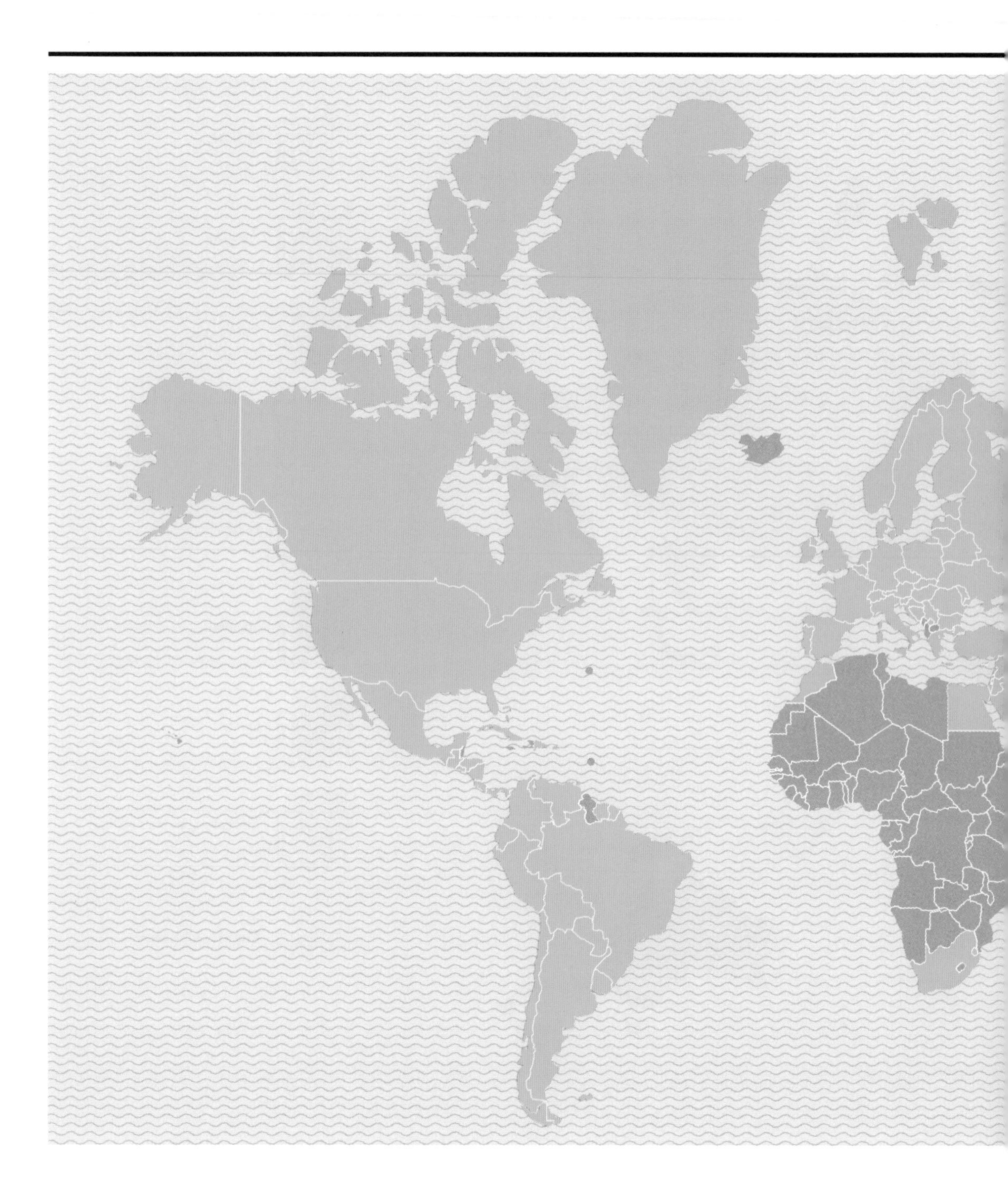

No McDonald's

McDonald's

35 Cats vs. dogs

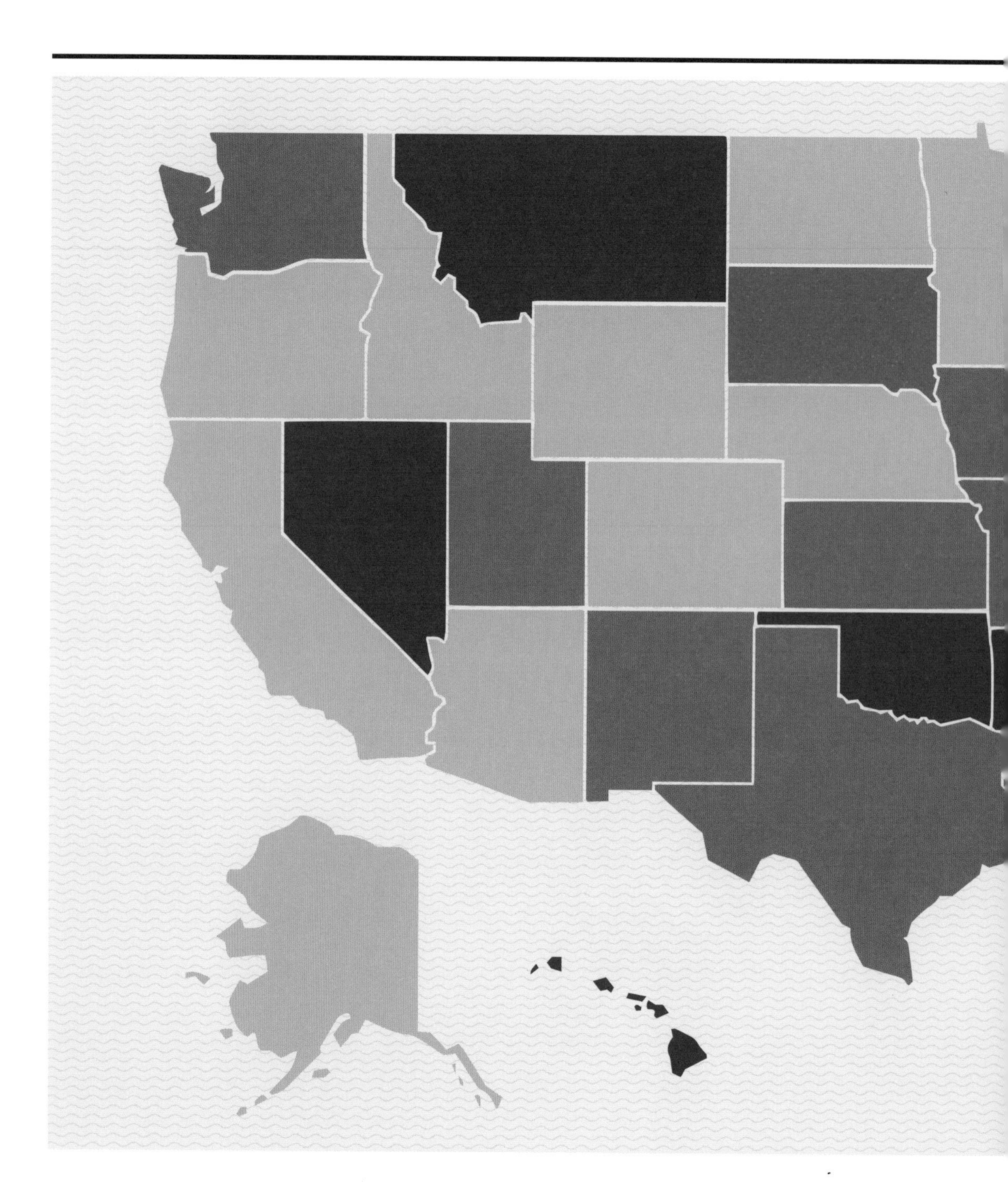

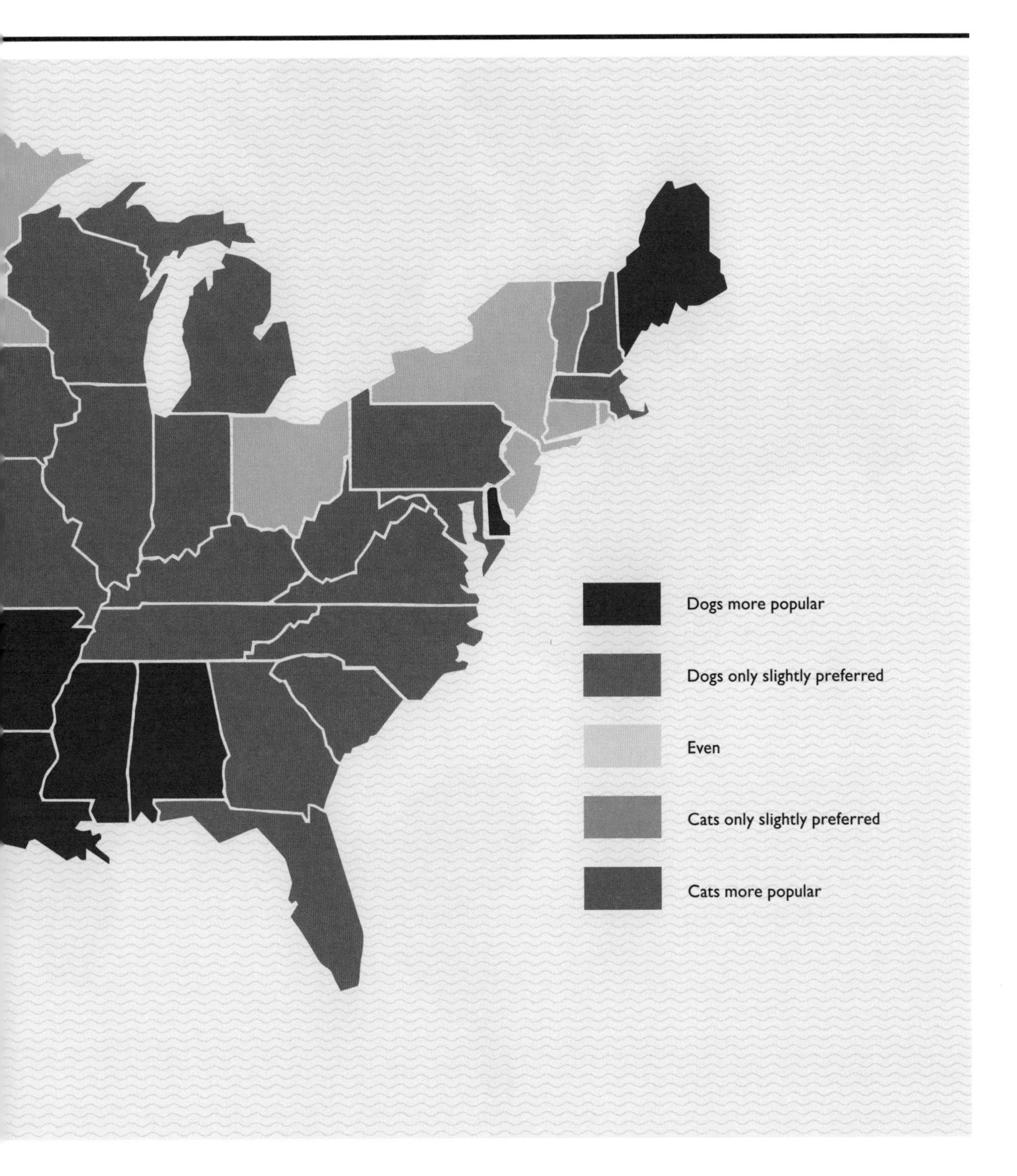
Dogs more popular
Dogs only slightly preferred
Even
Cats only slightly preferred
Cats more popular

36 The **most photographed places** in the world

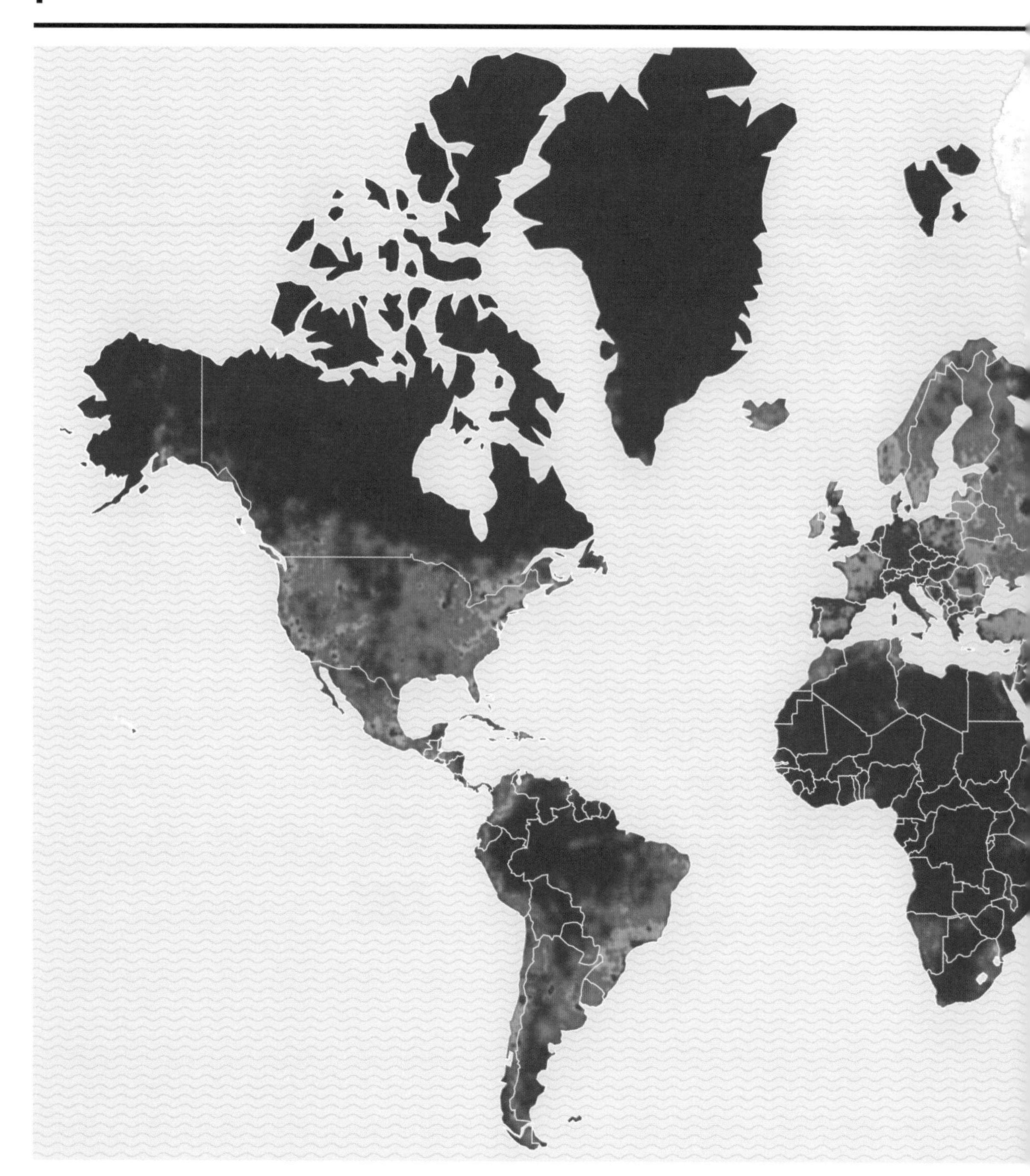

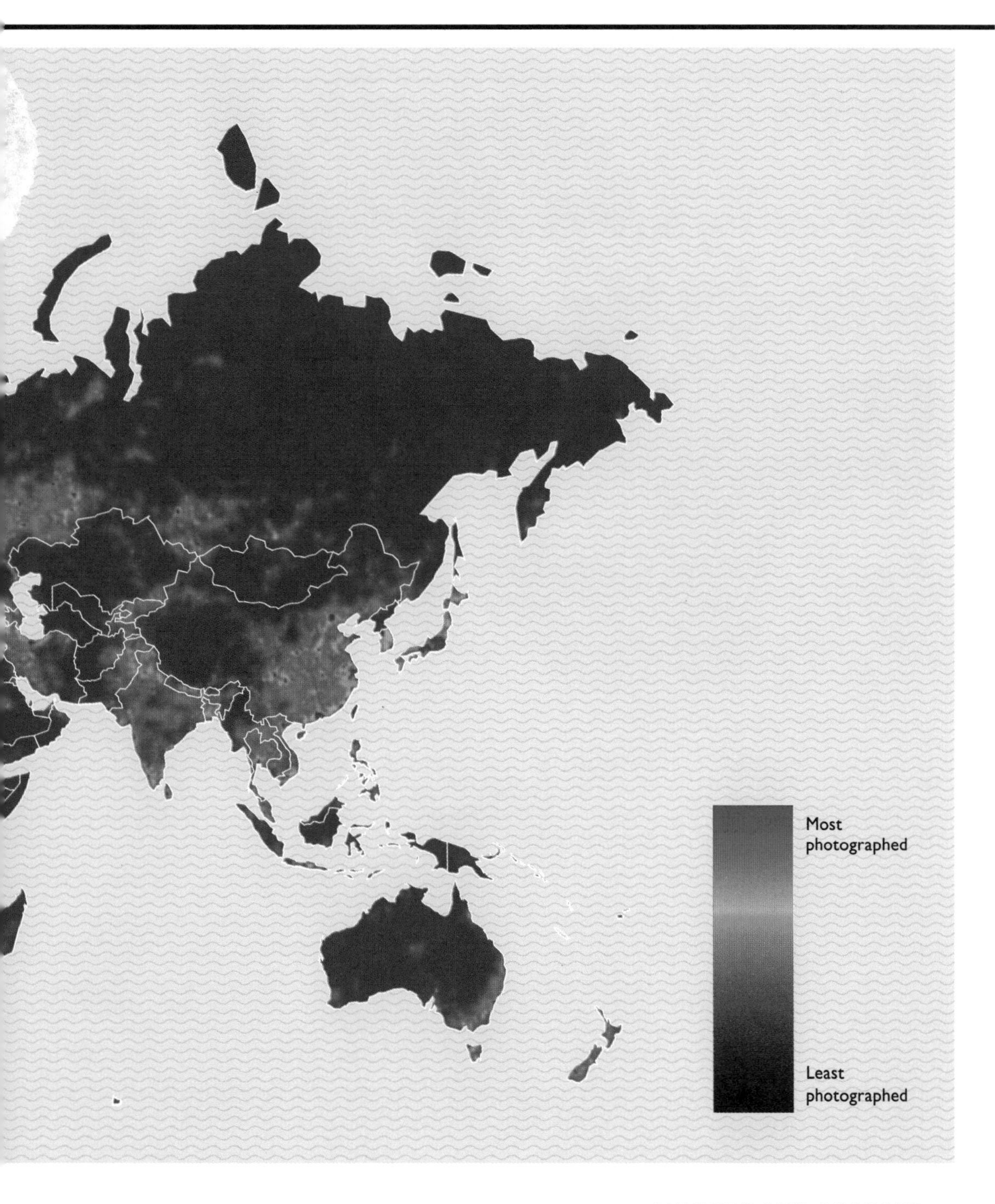
Most
photographed
Least
photographed

37 Heavy metal bands
per 100K people

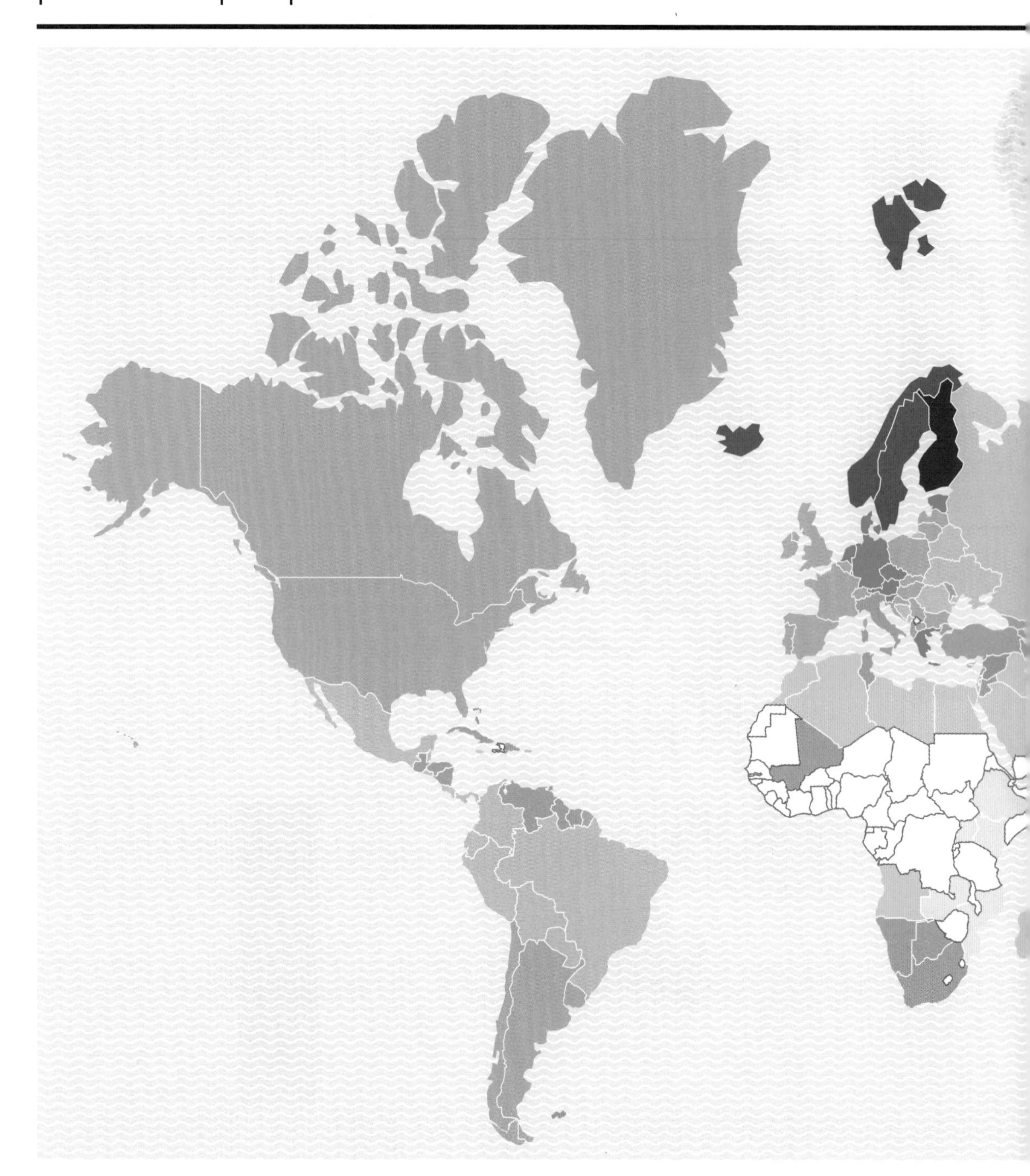

0

0.001

0.01

0.07

0.4

1.5

4.5

12

30

65.9

38 Countries with the most **Miss World winners**

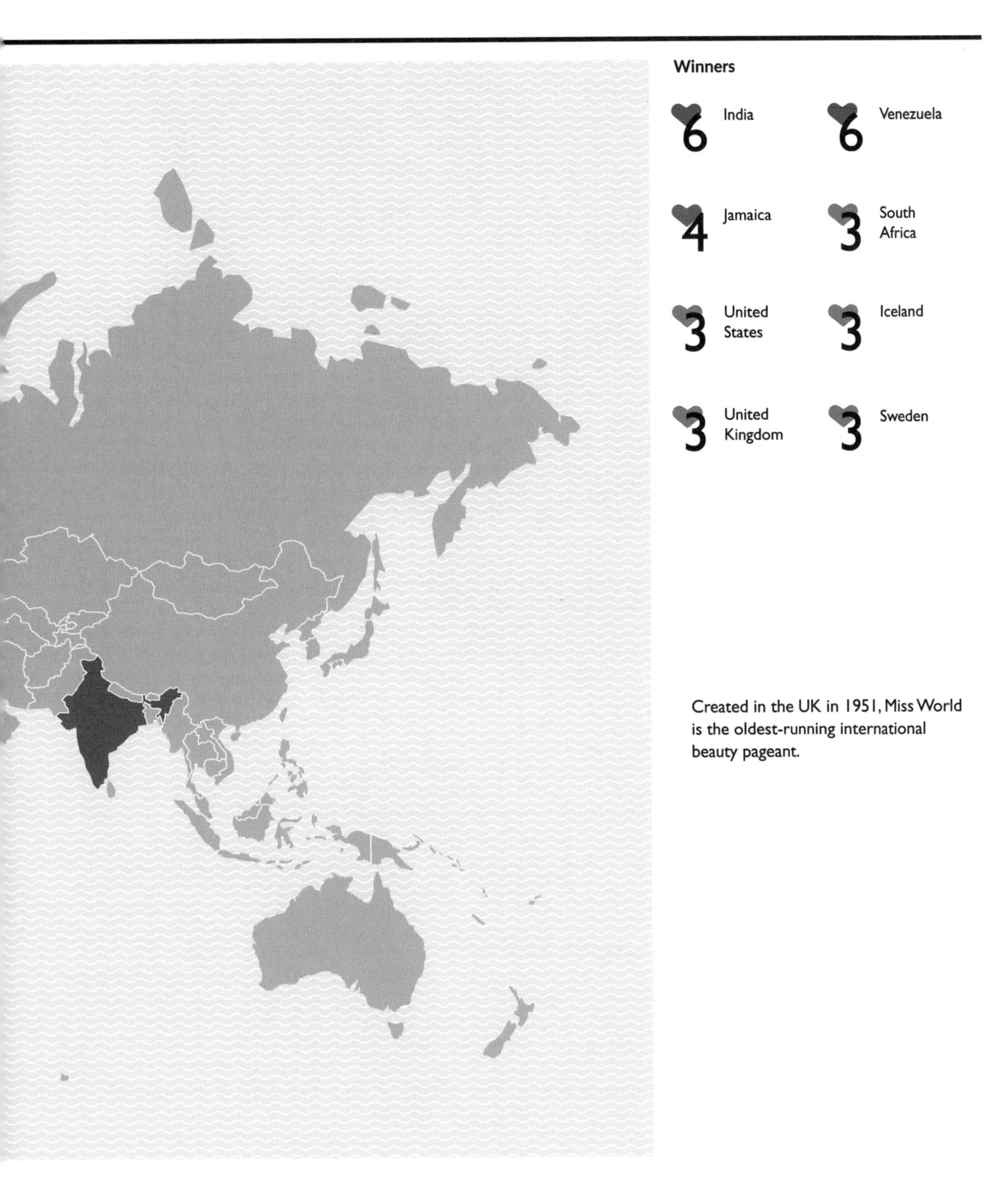

Created in the UK in 1951, Miss World is the oldest-running international beauty pageant.

39 Longest place names

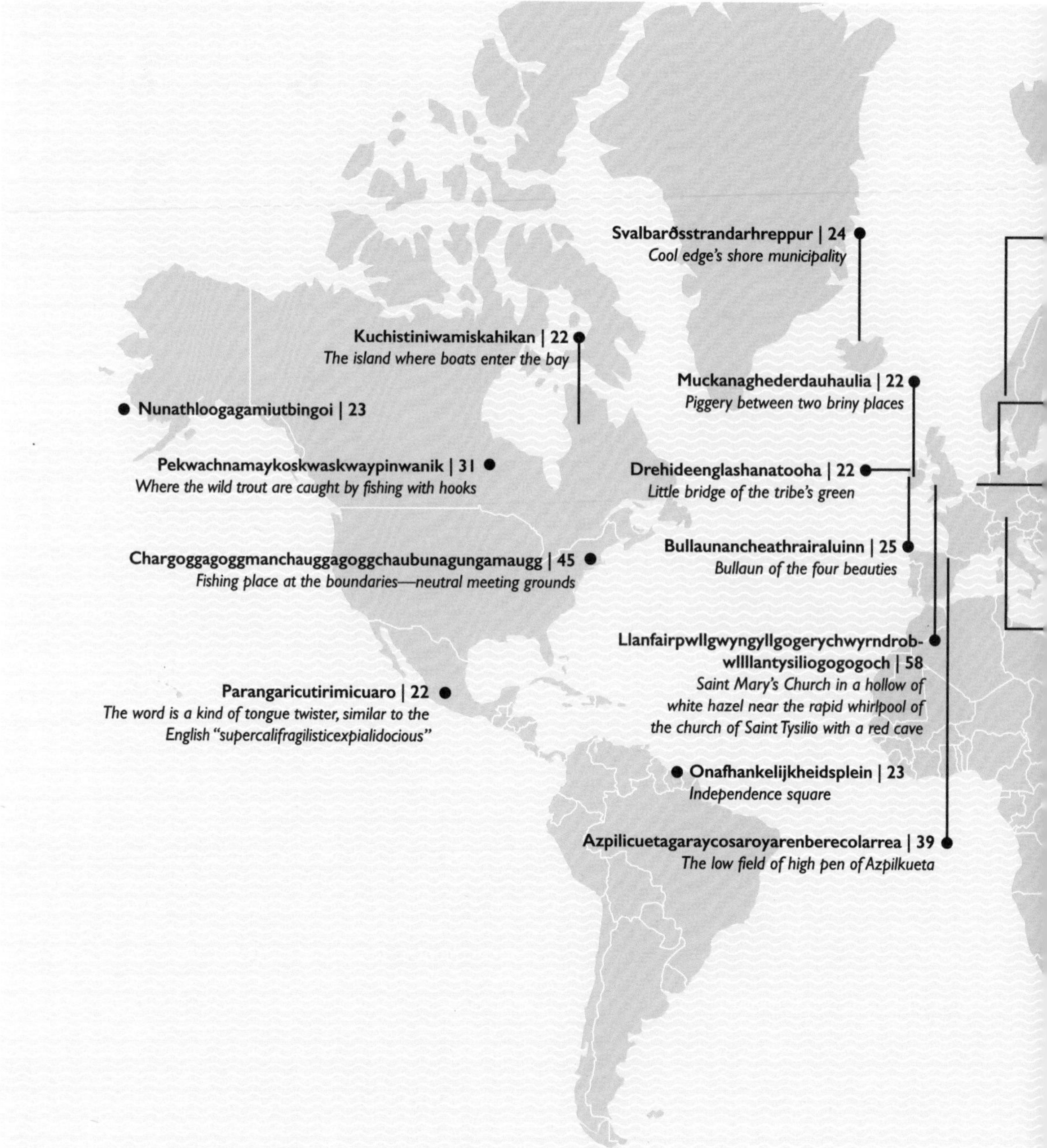

Svalbarðsstrandarhreppur | 24
Cool edge's shore municipality

Kuchistiniwamiskahikan | 22
The island where boats enter the bay

Nunathloogagamiutbingoi | 23

Muckanaghederdauhaulia | 22
Piggery between two briny places

Pekwachnamaykoskwaskwaypinwanik | 31
Where the wild trout are caught by fishing with hooks

Drehideenglashanatooha | 22
Little bridge of the tribe's green

Chargoggagoggmanchauggagoggchaubunagungamaugg | 45
Fishing place at the boundaries—neutral meeting grounds

Bullaunancheathrairaluinn | 25
Bullaun of the four beauties

Llanfairpwllgwyngyllgogerychwyrndrob-wllllantysiliogogogoch | 58
Saint Mary's Church in a hollow of white hazel near the rapid whirlpool of the church of Saint Tysilio with a red cave

Parangaricutirimicuaro | 22
The word is a kind of tongue twister, similar to the English "supercalifragilisticexpialidocious"

Onafhankelijkheidsplein | 23
Independence square

Azpilicuetagaraycosaroyarenberecolarrea | 39
The low field of high pen of Azpilkueta

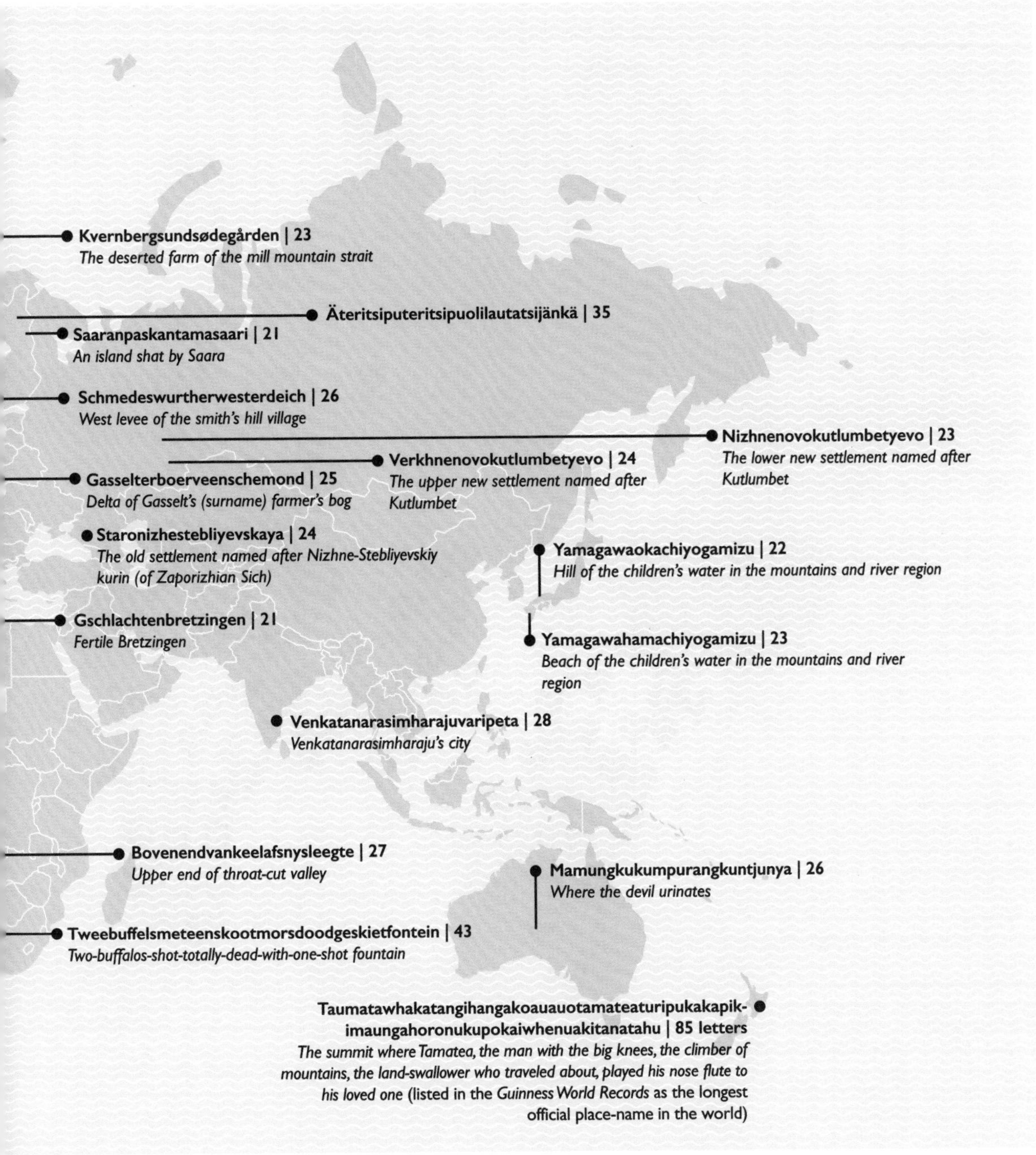

Kvernbergsundsødegården | 23
The deserted farm of the mill mountain strait

Äteritsiputeritsipuolilautatsijänkä | 35

Saaranpaskantamasaari | 21
An island shat by Saara

Schmedeswurtherwesterdeich | 26
West levee of the smith's hill village

Nizhnenovokutlumbetyevo | 23
The lower new settlement named after Kutlumbet

Verkhnenovokutlumbetyevo | 24
The upper new settlement named after Kutlumbet

Gasselterboerveenschemond | 25
Delta of Gasselt's (surname) farmer's bog

Staronizhestebliyevskaya | 24
The old settlement named after Nizhne-Stebliyevskiy kurin (of Zaporizhian Sich)

Yamagawaokachiyogamizu | 22
Hill of the children's water in the mountains and river region

Gschlachtenbretzingen | 21
Fertile Bretzingen

Yamagawahamachiyogamizu | 23
Beach of the children's water in the mountains and river region

Venkatanarasimharajuvaripeta | 28
Venkatanarasimharaju's city

Bovenendvankeelafsnysleegte | 27
Upper end of throat-cut valley

Mamungkukumpurangkuntjunya | 26
Where the devil urinates

Tweebuffelsmeteenskootmorsdoodgeskietfontein | 43
Two-buffalos-shot-totally-dead-with-one-shot fountain

Taumatawhakatangihangakoauauotamateaturipukakapik-imaungahoronukupokaiwhenuakitanatahu | 85 letters
The summit where Tamatea, the man with the big knees, the climber of mountains, the land-swallower who traveled about, played his nose flute to his loved one (listed in the *Guinness World Records* as the longest official place-name in the world)

40 Most **recurring word*** on each country's English Wikipedia page

* Excluding "country," linking words, demonyms, and "government"

1. Denmark
2. Quebec
3. War
4. Indigenous
5. Island
6. States
7. British
8. City
9. National
10. Santo Domingo
11. Central
12. Canal
13. Island
14. Oil
15. Department
16. World
17. Dutch
18. Peru
19. Spanish
20. La Paz
21. Portuguese
22. Population
23. South
24. Montevideo
25. Buenos Aires
26. Reykjavik
27, 28, 29. World
30. Prague
31. Ireland
32. State
33. World
34. Soviet
35, 36. Baltic
37. Soviet
38. Century
39. Federal
40. European
41. German
42. Bratislava
43. Soviet
44, 45. World
46. Soviet
47. Sarajevo
48. War
49. Lisbon
50. World
51. Sahara
52. Athens
53. French
54. Ottoman
55. Arab
56. World
57. Gaddafi
58. Province
59. Maroc
60. Ould
61. Population
62. Military
63. President
64. World
65. King
66. Forces
67. Dubai
68. Sultan
69. Sana'a
70. Ethiopia
71. South
72. Africa
73. President
74. River
75. Mining
76, 77. French
78. French
79. President
80. Southern
81. Population
82. Federal
83. Years
84. National
85. President
86. Republic
87. National
88. Portuguese
89. Rhodesia
90. President
91. Zanzibar
92. Island
93. Shona
94, 95. Africa
96. Cape
97. South
98. Portuguese
99. Russian
100. Soviet
101. Baku
102, 103. Soviet
104. Niyazov
105. Soviet
106. Kabul
107. Persian
108. World
109. Dynasty
110. Population
111. Soviet
112. South
113. North
114. World
115, 116. India
117. Bengal
118. South
119. Military
120. French
121. Bangkok
122. Hmong
123. Khmer
124. Tamil
125. State
126. Region
127. Java
128. Portuguese
129. National
130. Guadalcanal
131. New
132. Island

41 Age of consent
for heterosexual sex

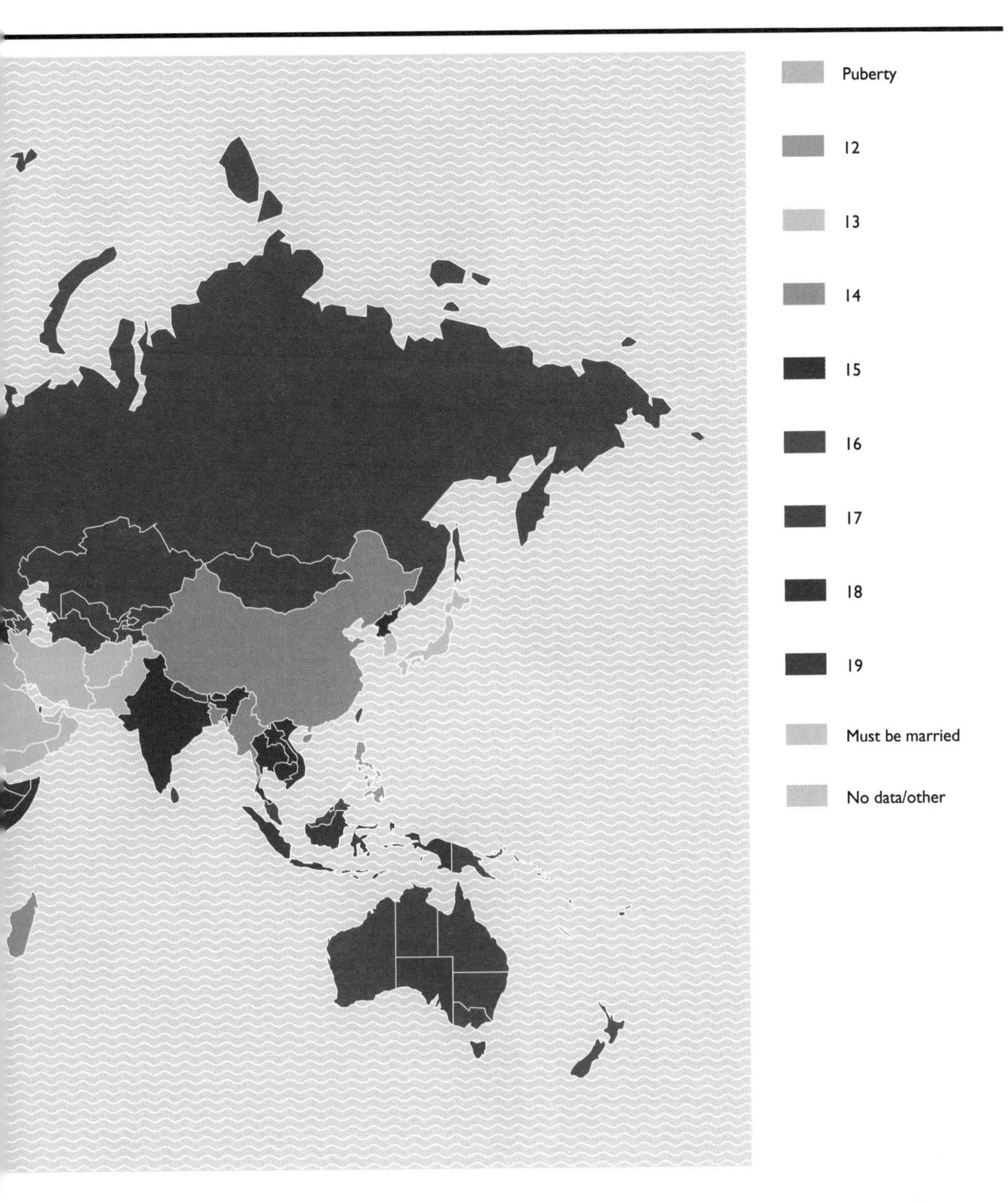
Puberty
12
13
14
15
16
17
18
19
Must be married
No data/other

42 **Male circumcision:** one thing that unites the US and the Middle East

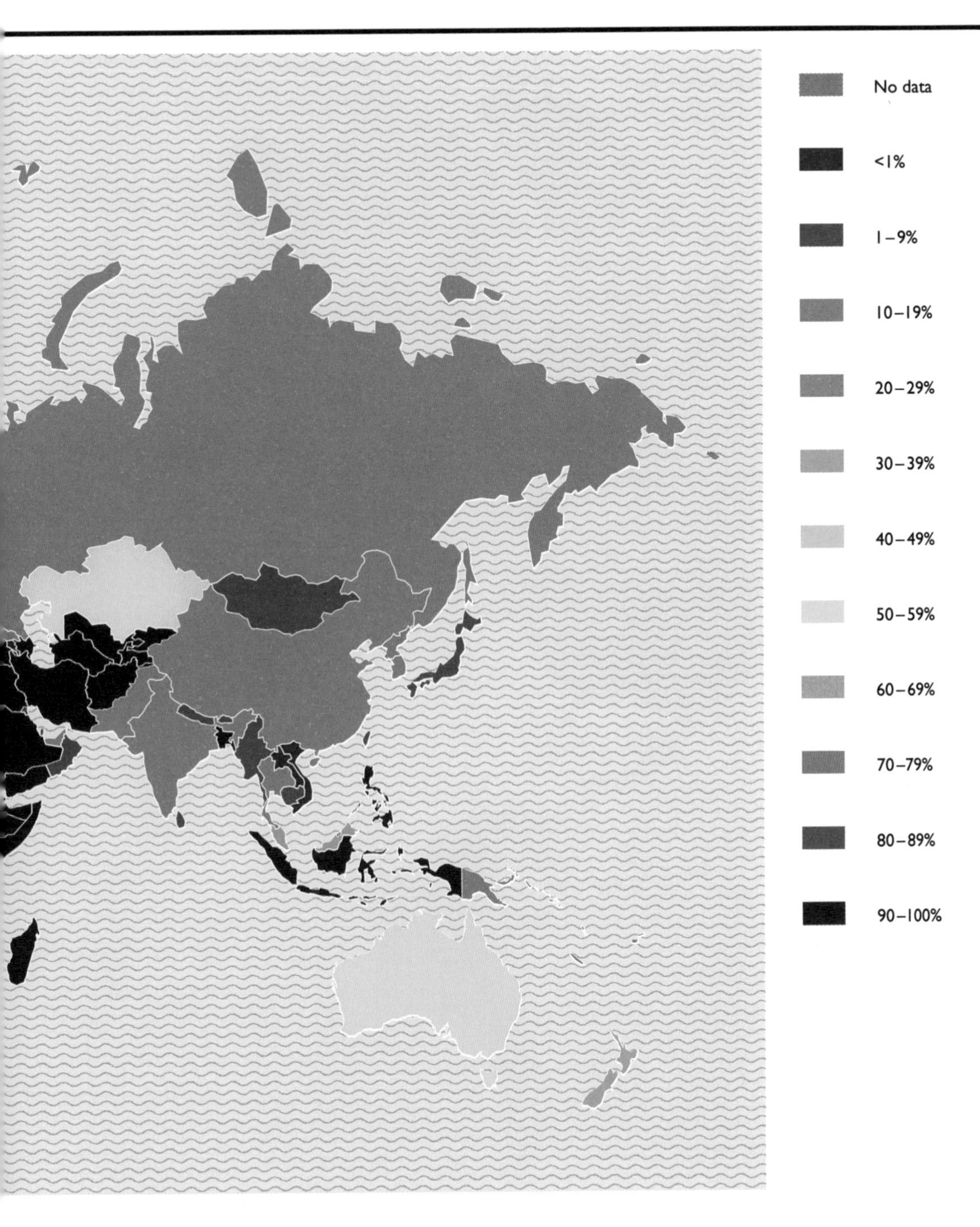

No data
<1%
1–9%
10–19%
20–29%
30–39%
40–49%
50–59%
60–69%
70–79%
80–89%
90–100%

43 World **plug and socket** map

Type A
Type B
Type C
Type E
Type F
Type D
Type M
Type N
Type G
Type H
Type I
Type J
Type K
Type L

4

FRIENDS AND ENEMIES

44 **Open borders** of the world*

* In 2018

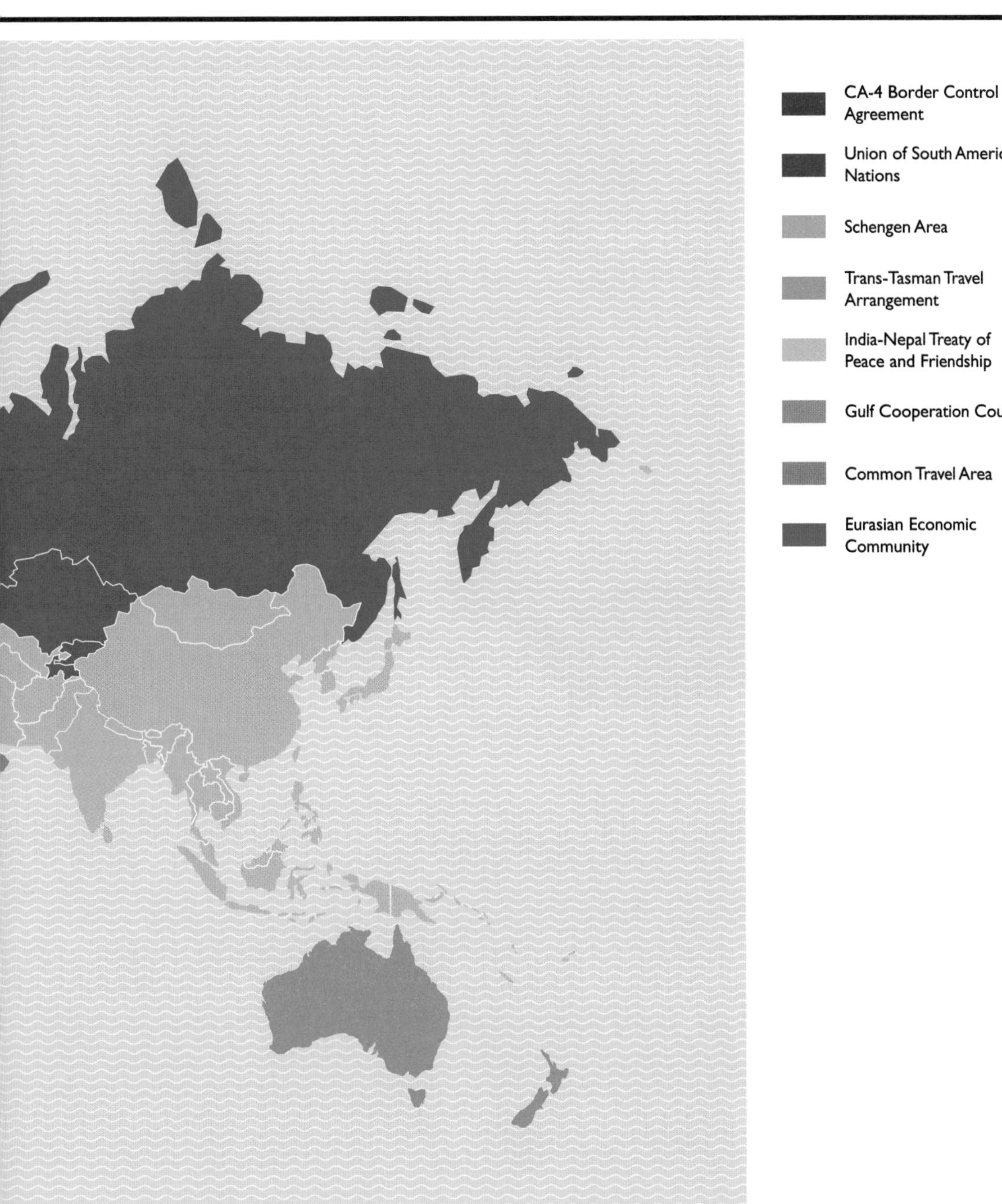
CA-4 Border Control Agreement
Union of South American Nations
Schengen Area
Trans-Tasman Travel Arrangement
India-Nepal Treaty of Peace and Friendship
Gulf Cooperation Council
Common Travel Area
Eurasian Economic Community

45 Who Americans consider their **allies, friends, and enemies**

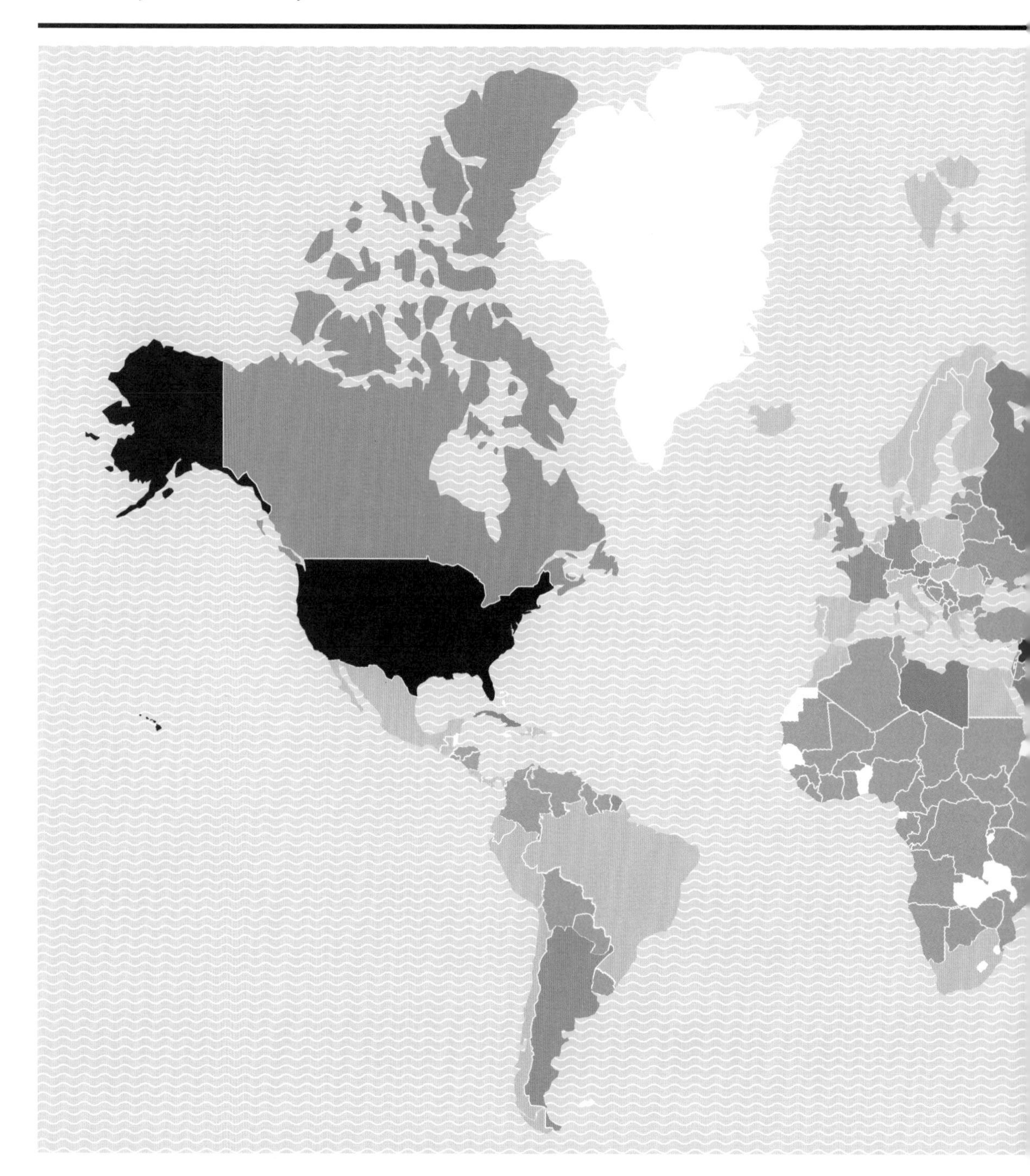

Ally

Friendly

Unfriendly

Enemy

Not sure

United States of America

No data

46 Countries the US is **obligated to go to war** for (for now)

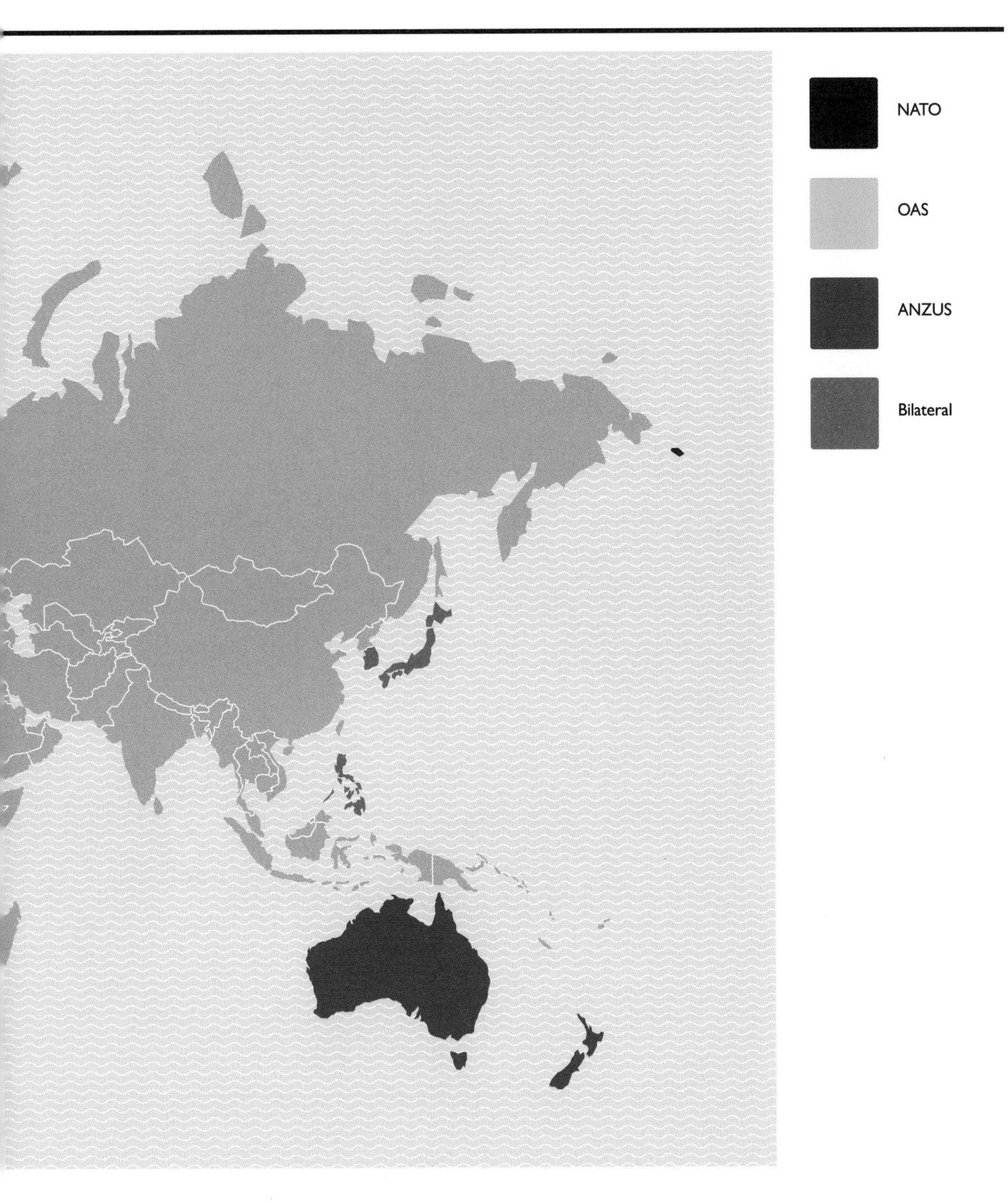
NATO
OAS
ANZUS
Bilateral

47 22 countries that the UK **has not invaded**

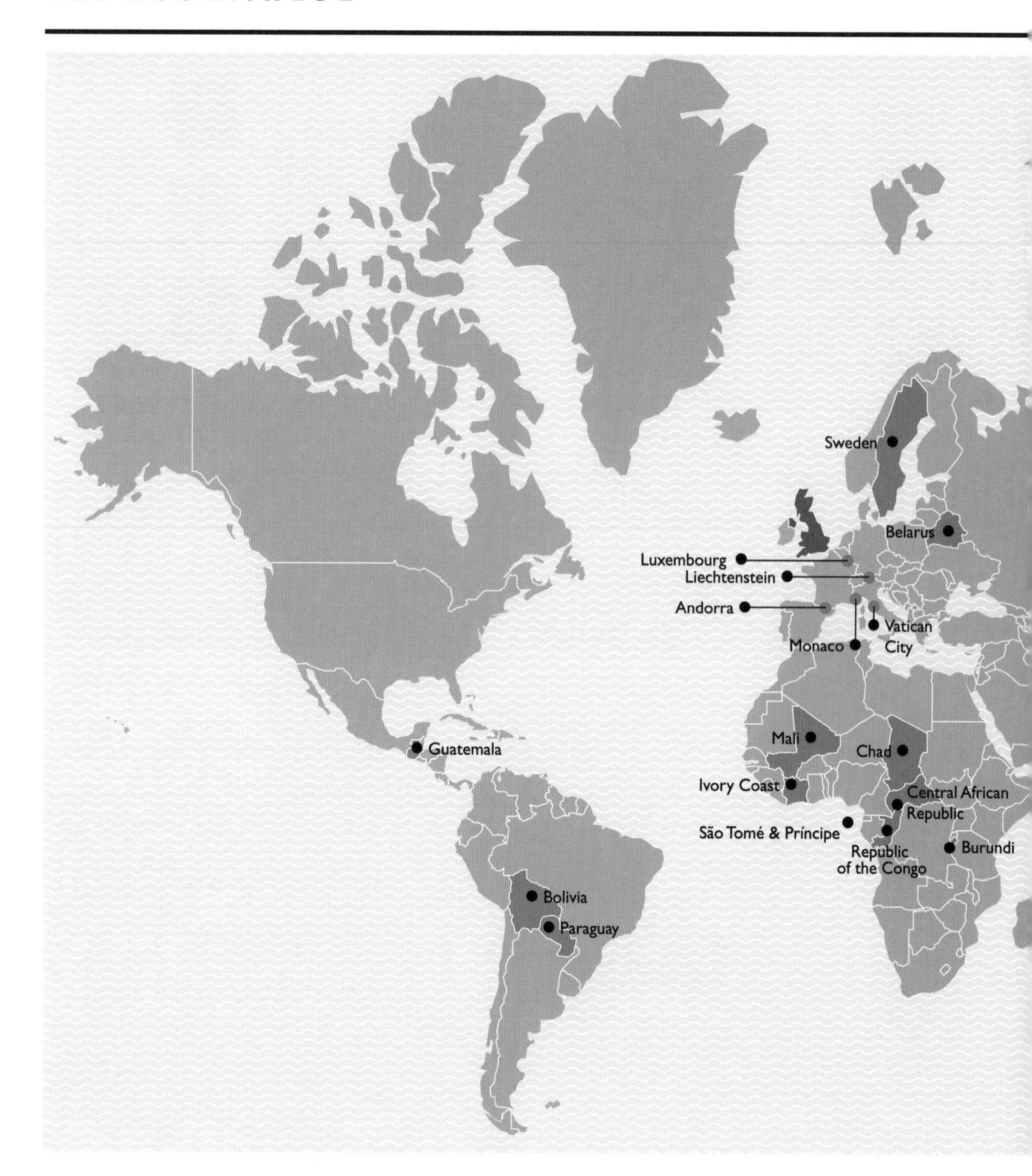

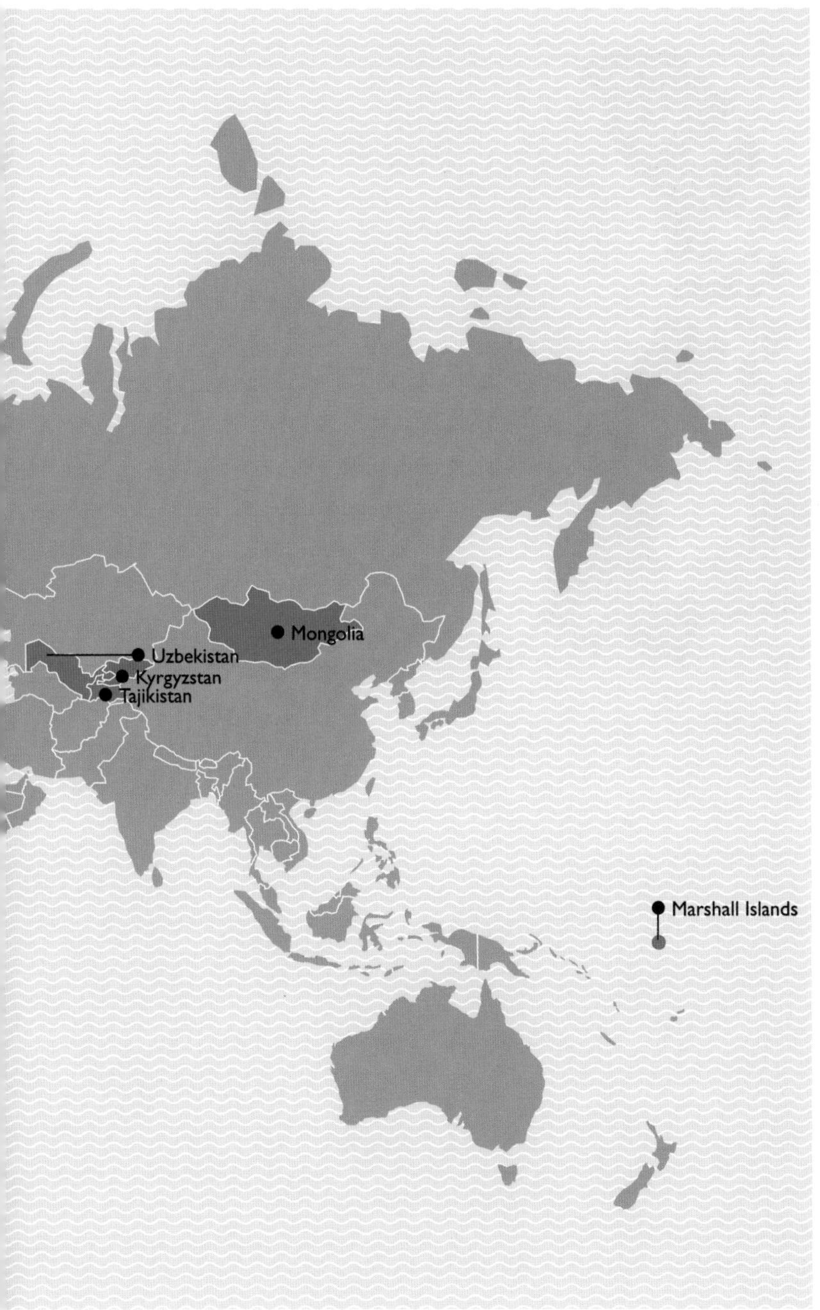

Countries not invaded or occupied by the UK

The UK

In grey are the countries (or their historical predecessors) who were fought or occupied by the UK (or its predecessors) at some point in history.

48 Countries that were raided or settled by **the Vikings***

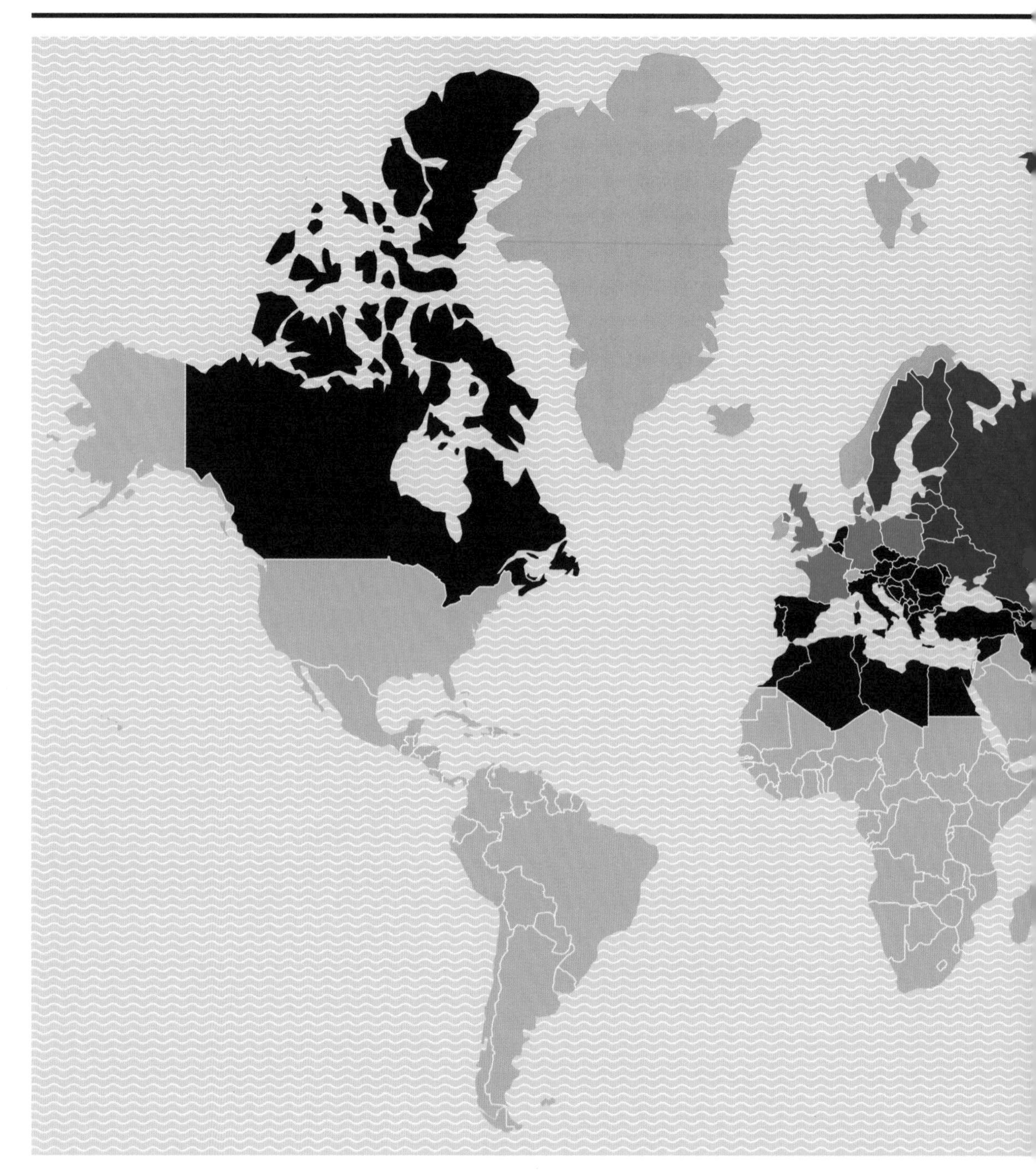

* Based on modern borders

Raided by Norse Vikings

Norwegian Viking raids and settlements

Danish Viking raids and settlements

Swedish Viking raids and settlements

49 European countries* that have **invaded Poland**

Poland

Invaders

Using modern political borders, this map shows invasions of the area we now know as Poland.

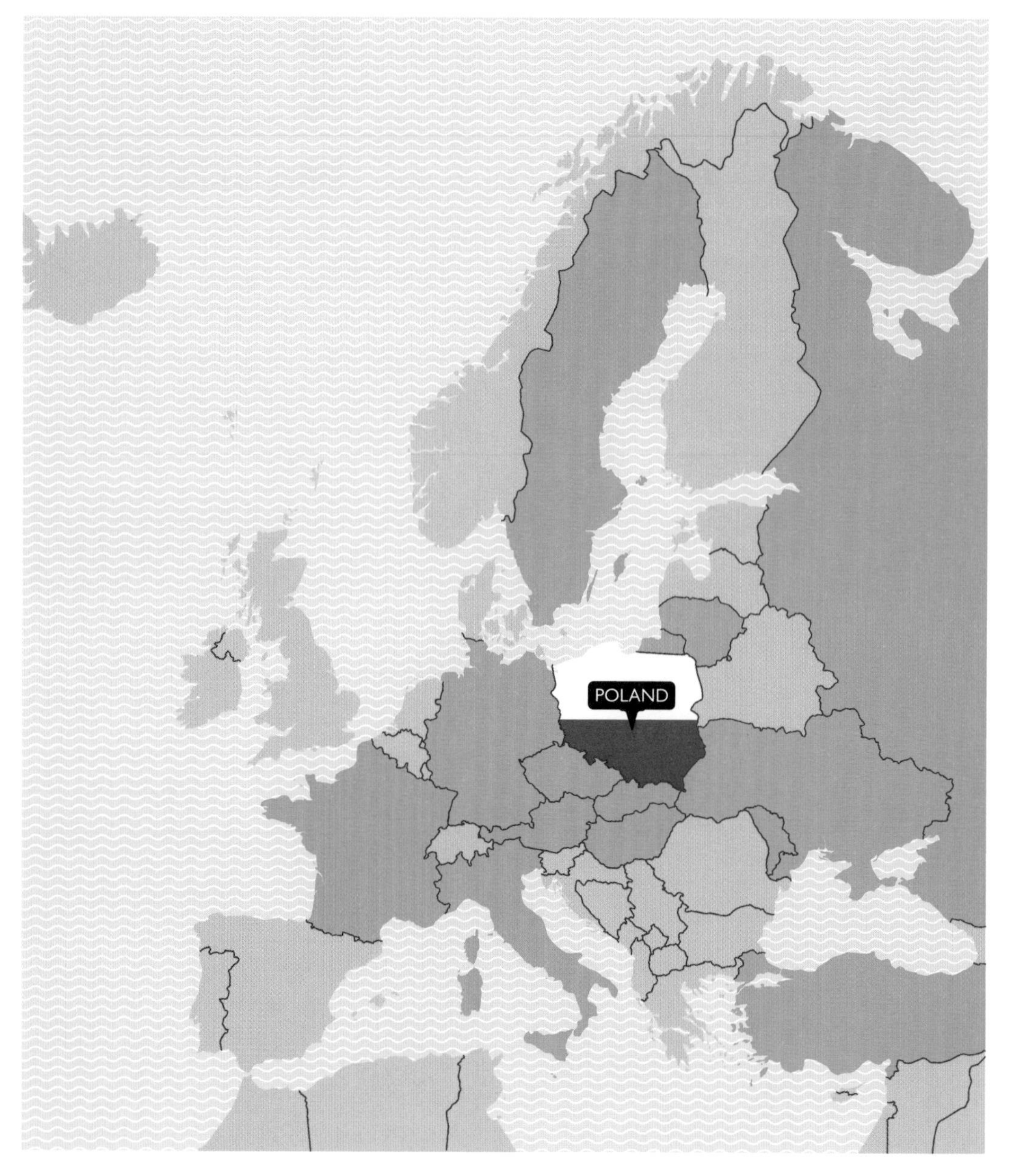

* Including Russia

50 **Zone Rouge:** an area of France so badly damaged during WW I that people were forbidden to live there*

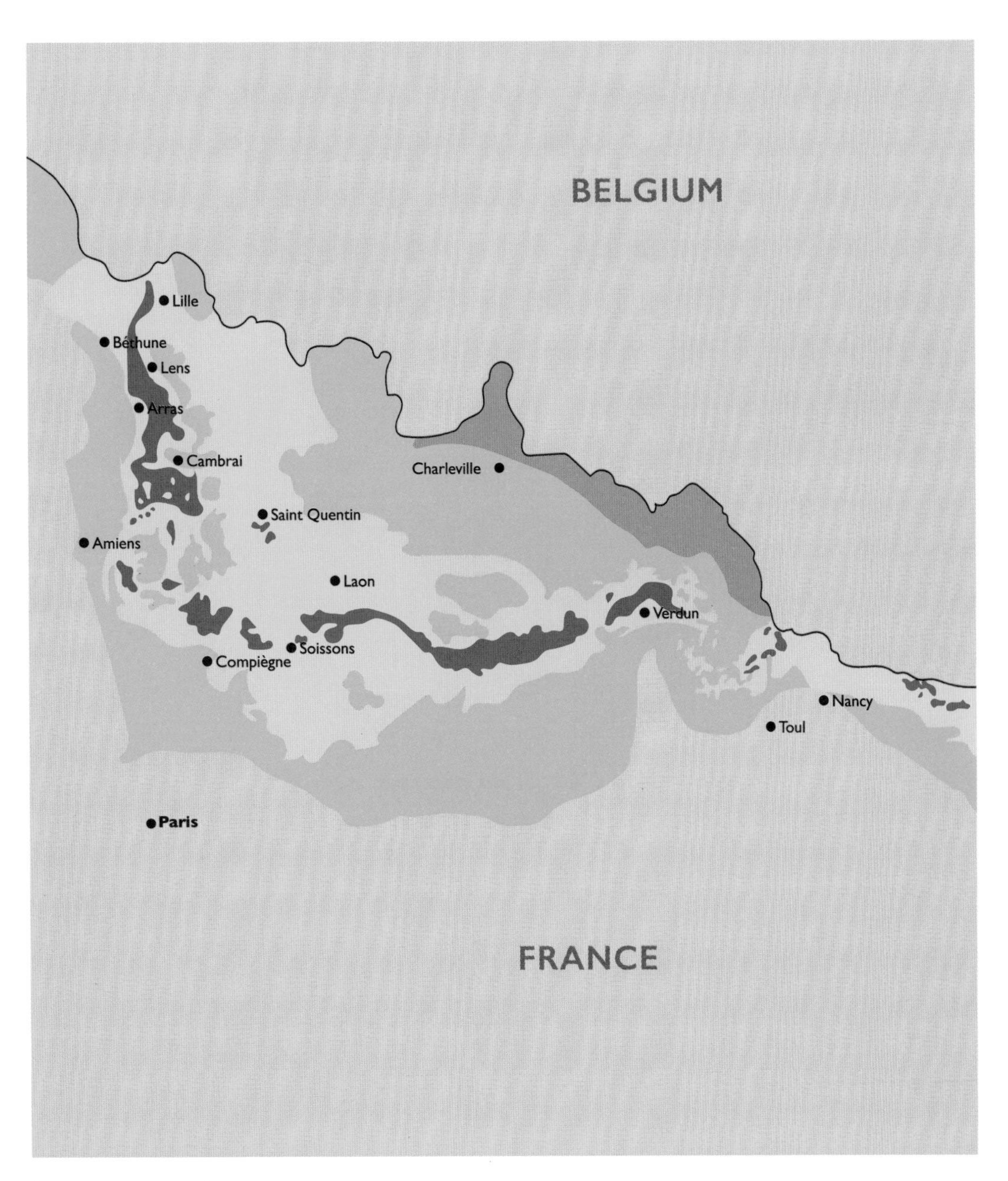

* Some areas remain off limits even to this day.

51 Countries that officially recognize the State of **Palestine**

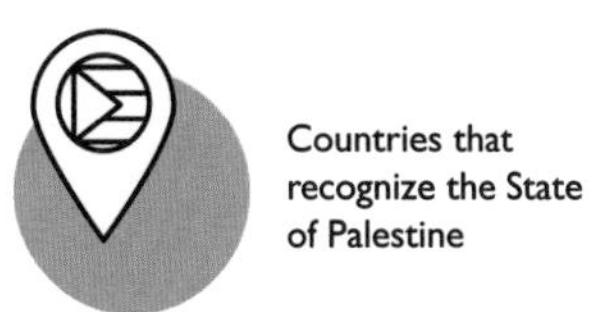
Countries that recognize the State of Palestine

52 Countries that officially recognize the State of **Israel**

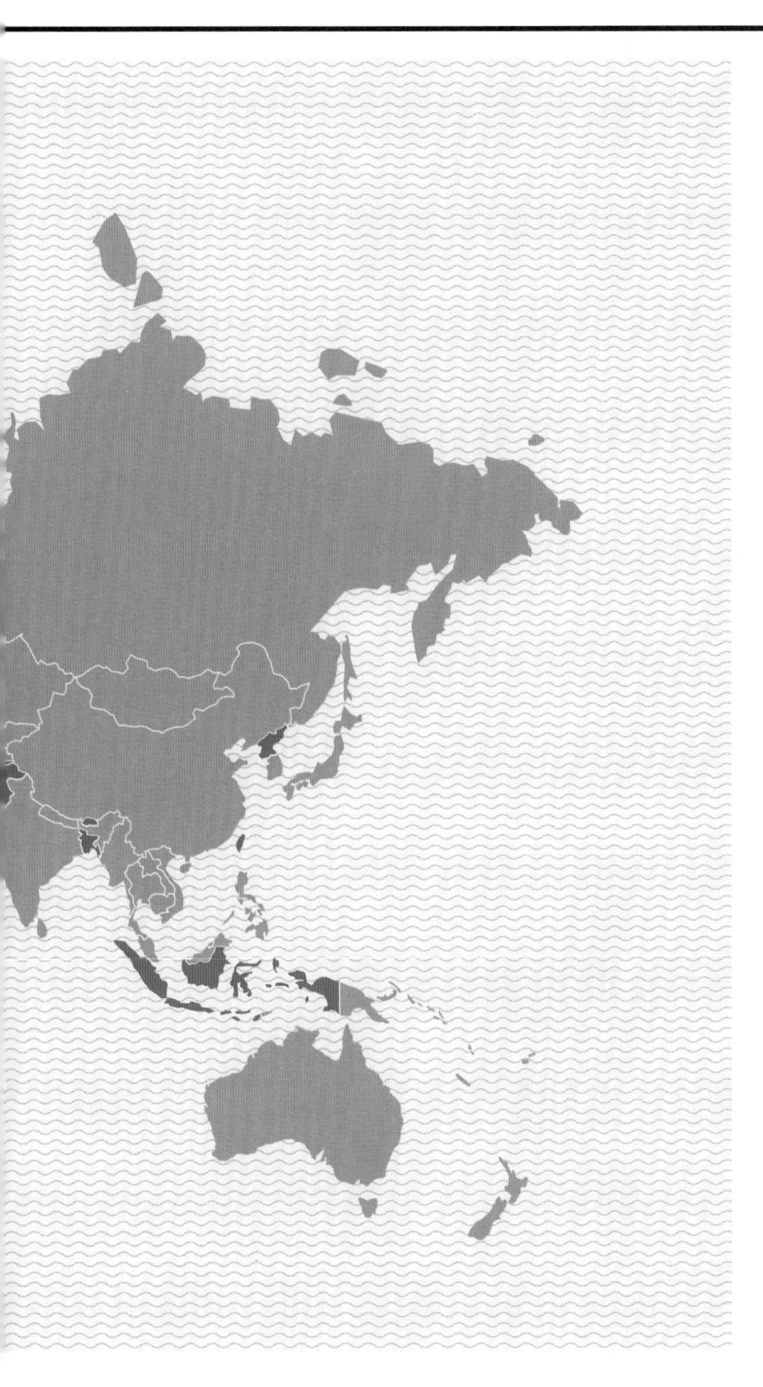

Diplomatic relations

Former diplomatic relations

No diplomatic relations
but former trade relations

No diplomatic relations

53 Where **North Korea has embassies**

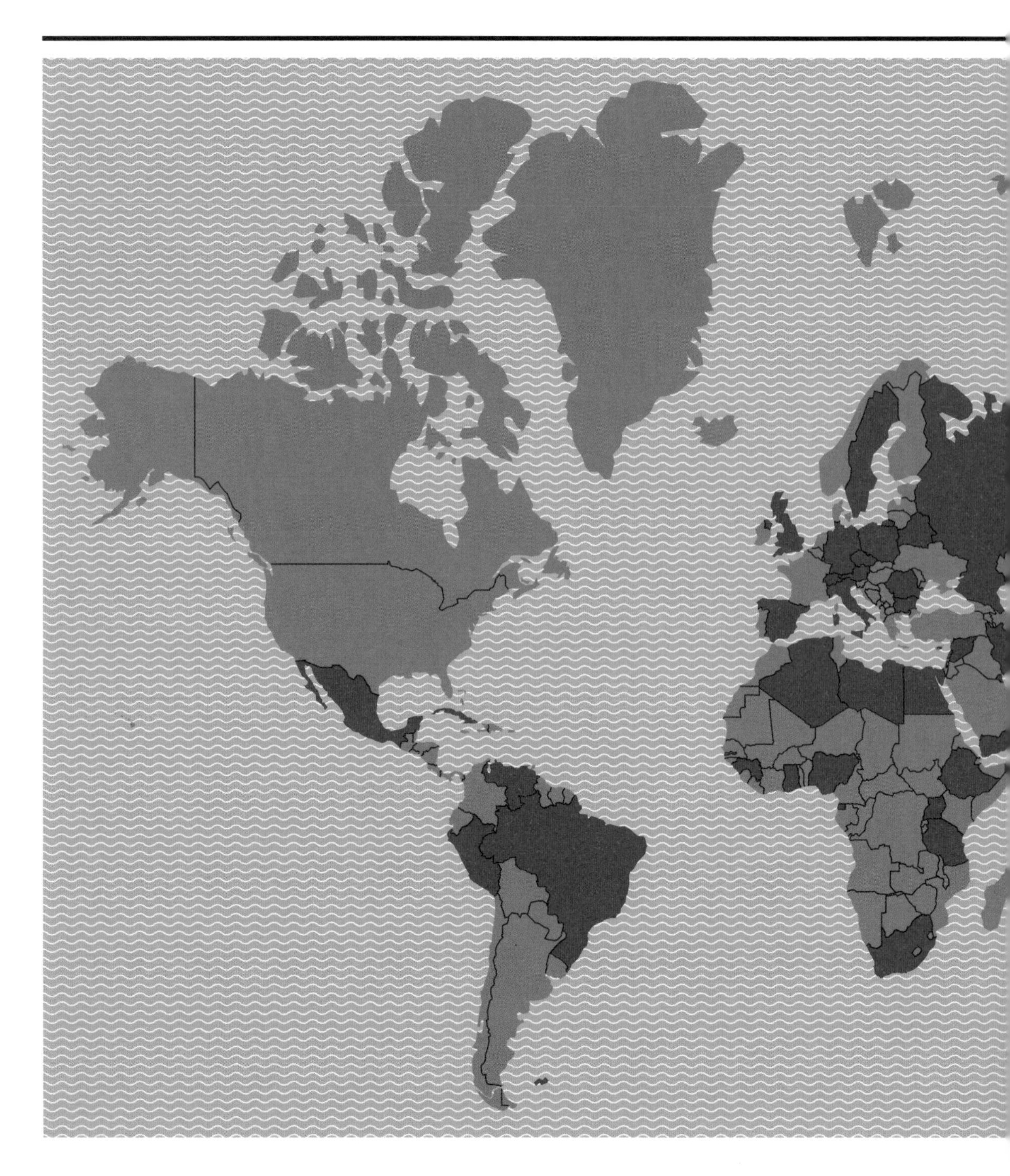

NORTH KOREA

Countries where North Korea has embassies

54 Who has **embassies in North Korea?**

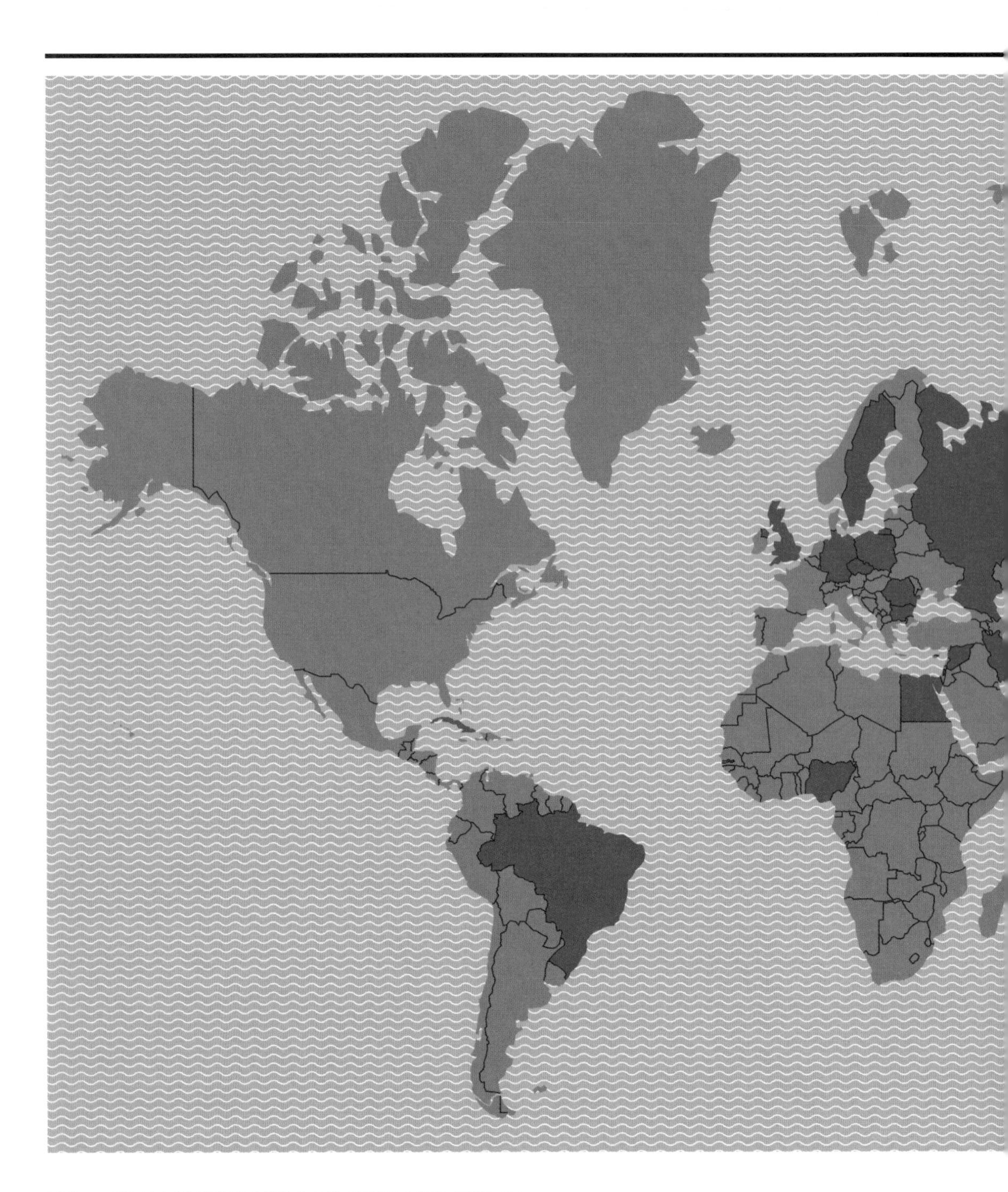

NORTH KOREA

Countries with
diplomatic missions
in North Korea

55 Former and current **communist** countries*

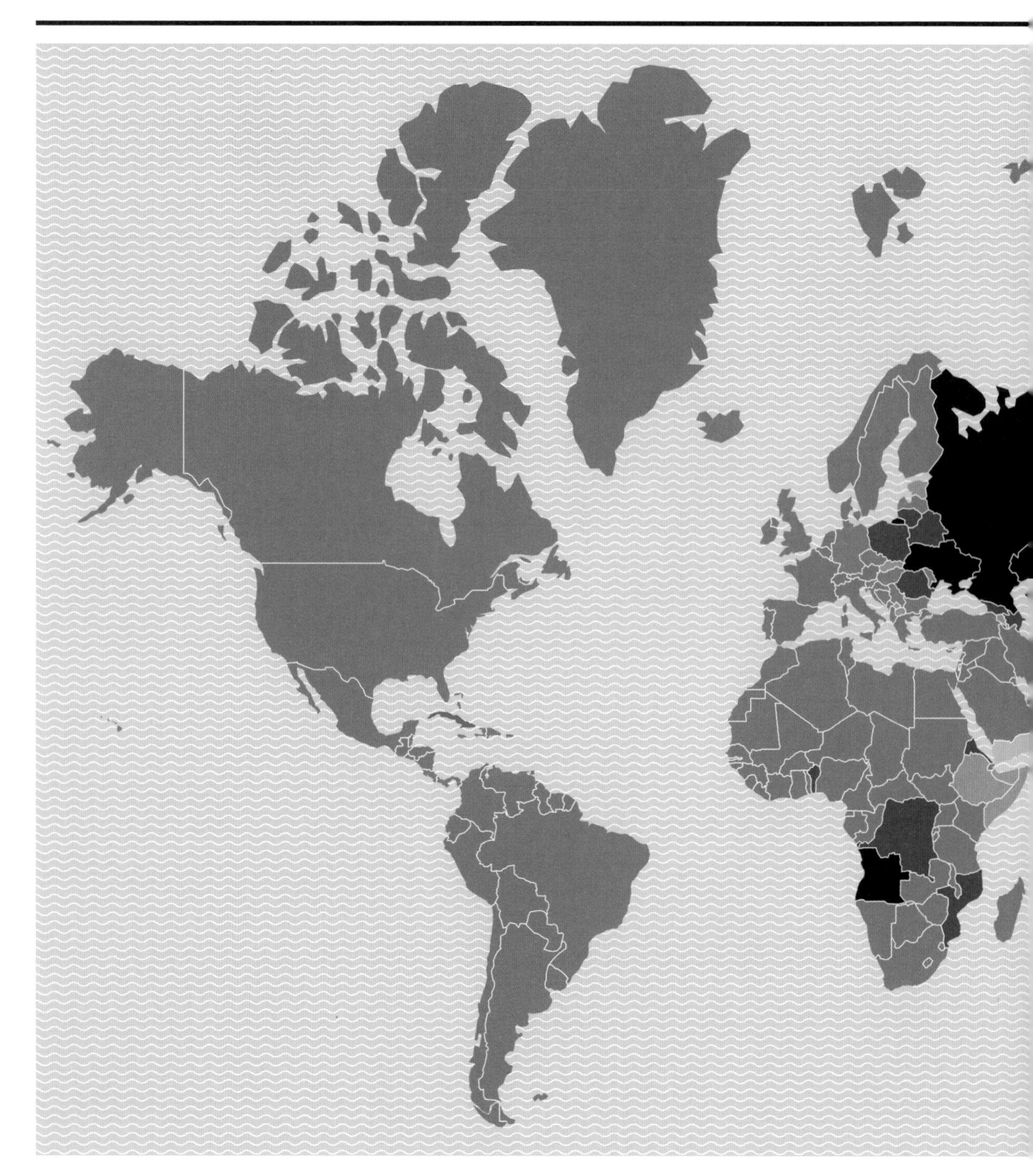

* In 2018

Presidential or semi-presidential constitutional republic

Presidential or semi-presidential republic

Parliamentary constitutional republic

Parliamentary republic

Constitutional monarchy

Islamic republic

Provisional government

Communist

5

GEOGRAPHY

56 Mercator projection vs. the true size of countries

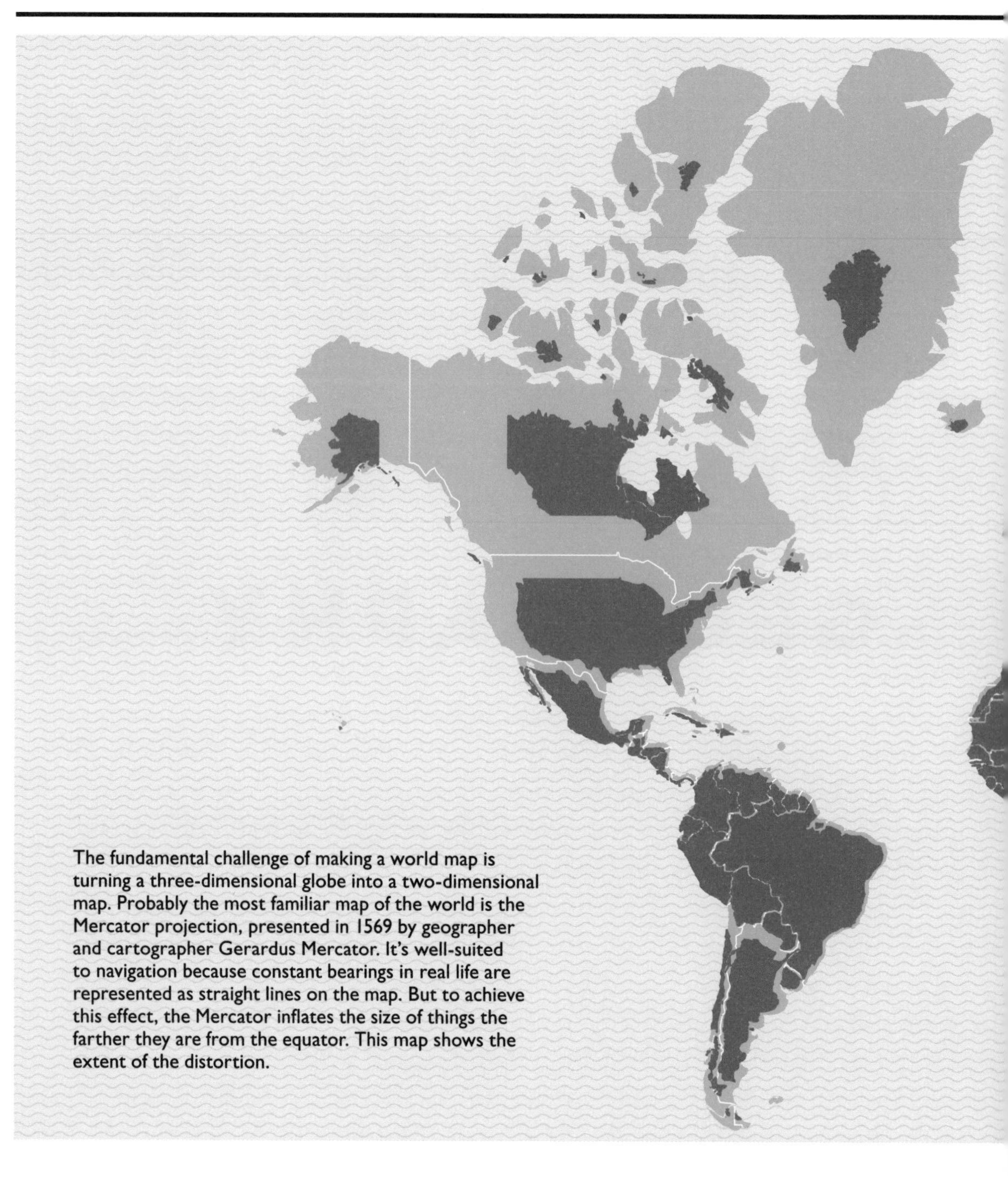

The fundamental challenge of making a world map is turning a three-dimensional globe into a two-dimensional map. Probably the most familiar map of the world is the Mercator projection, presented in 1569 by geographer and cartographer Gerardus Mercator. It's well-suited to navigation because constant bearings in real life are represented as straight lines on the map. But to achieve this effect, the Mercator inflates the size of things the farther they are from the equator. This map shows the extent of the distortion.

57 **Chile** is a ridiculously long country

You probably already know that Chile is a rather long (or tall) country, but I bet you don't know quite how long it is. From north to south, Chile extends **2,653** miles, yet is only **217** miles at its widest point and averages just **110** miles east to west.

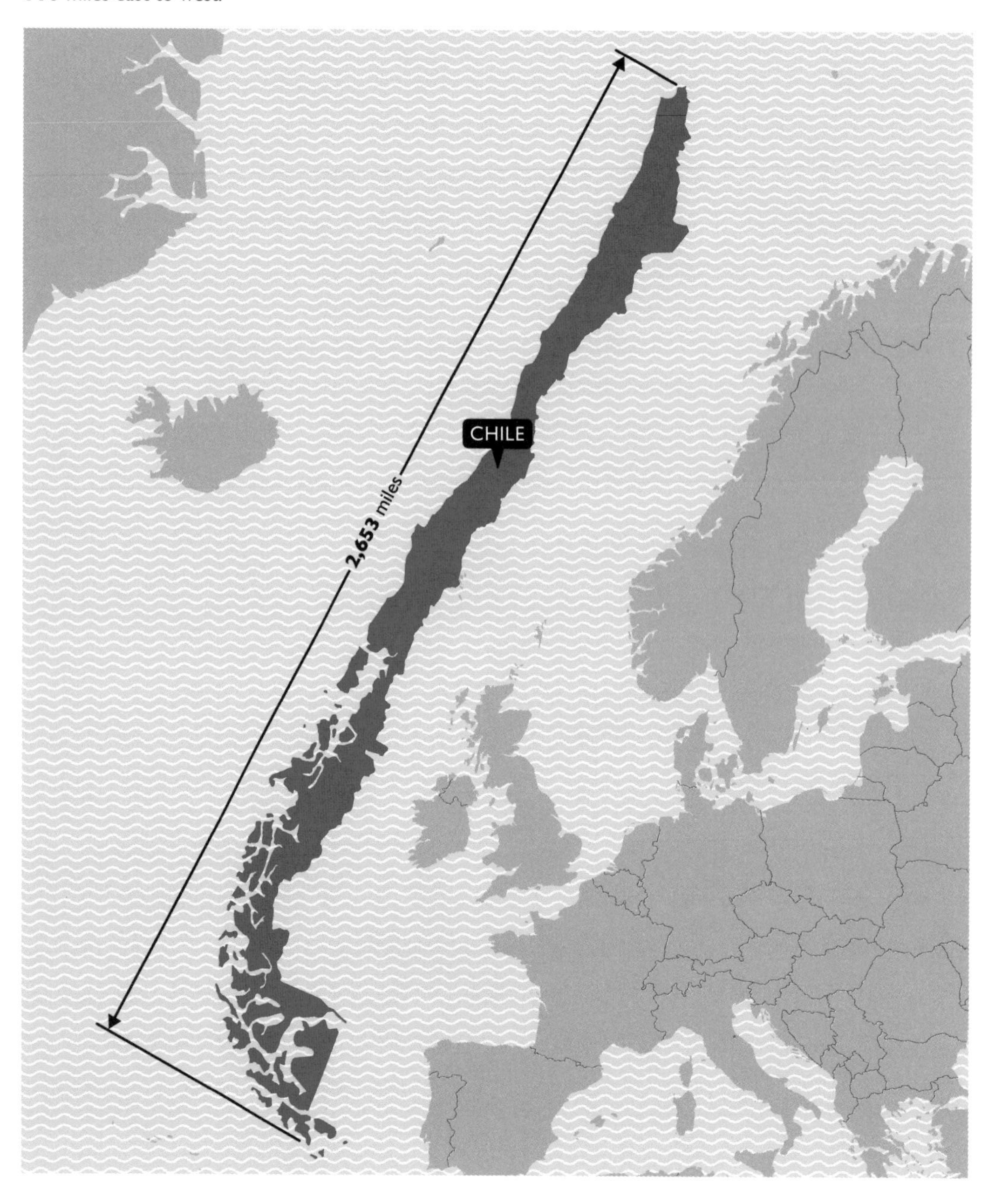

58 The true size of **Africa**

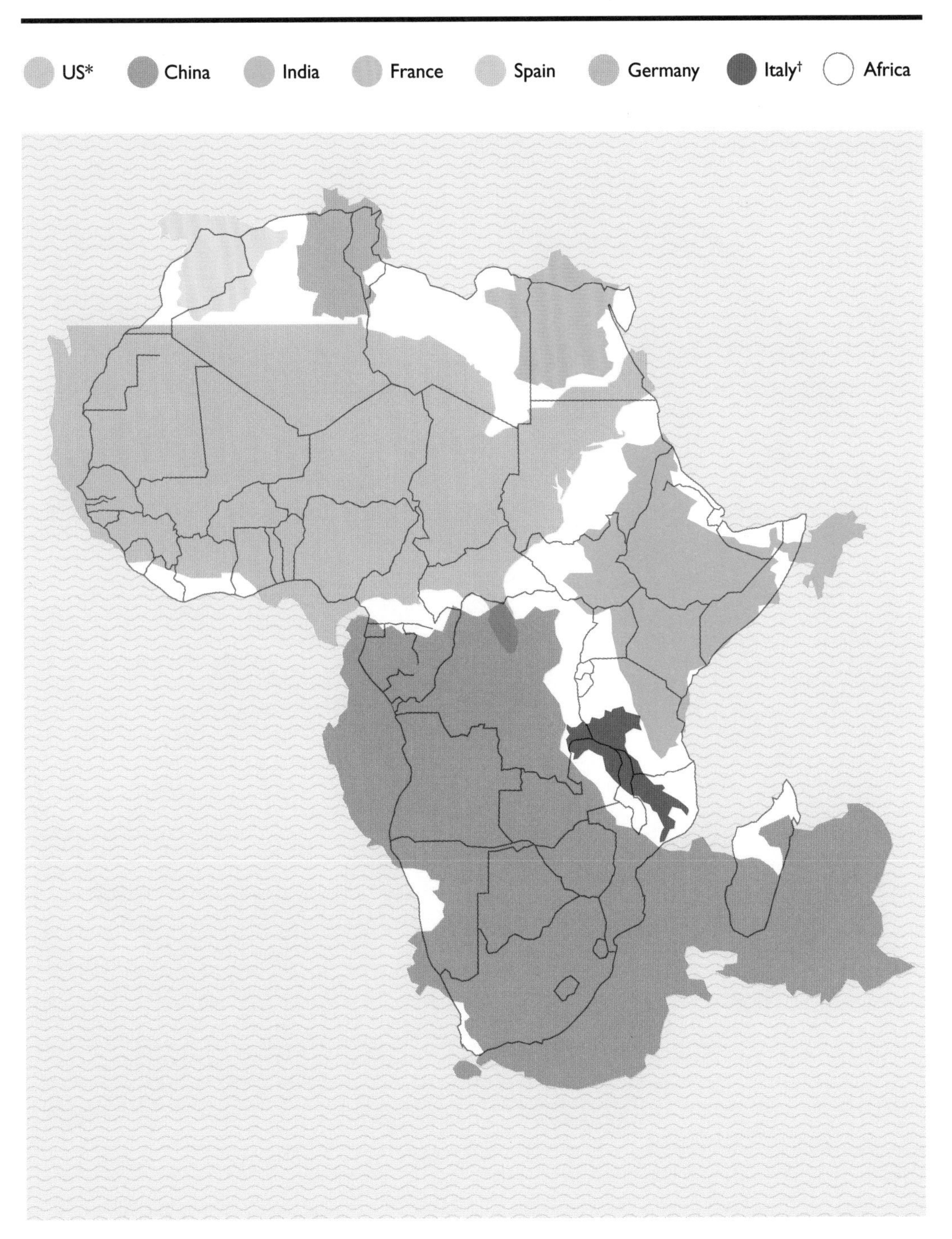

* Excluding Alaska and Hawaii

† Excluding Sardinia

59 The **Pacific Ocean** is larger than all the land on Earth

As difficult as it may be to believe, the Pacific Ocean is larger than the landmass of every single continent and island combined.

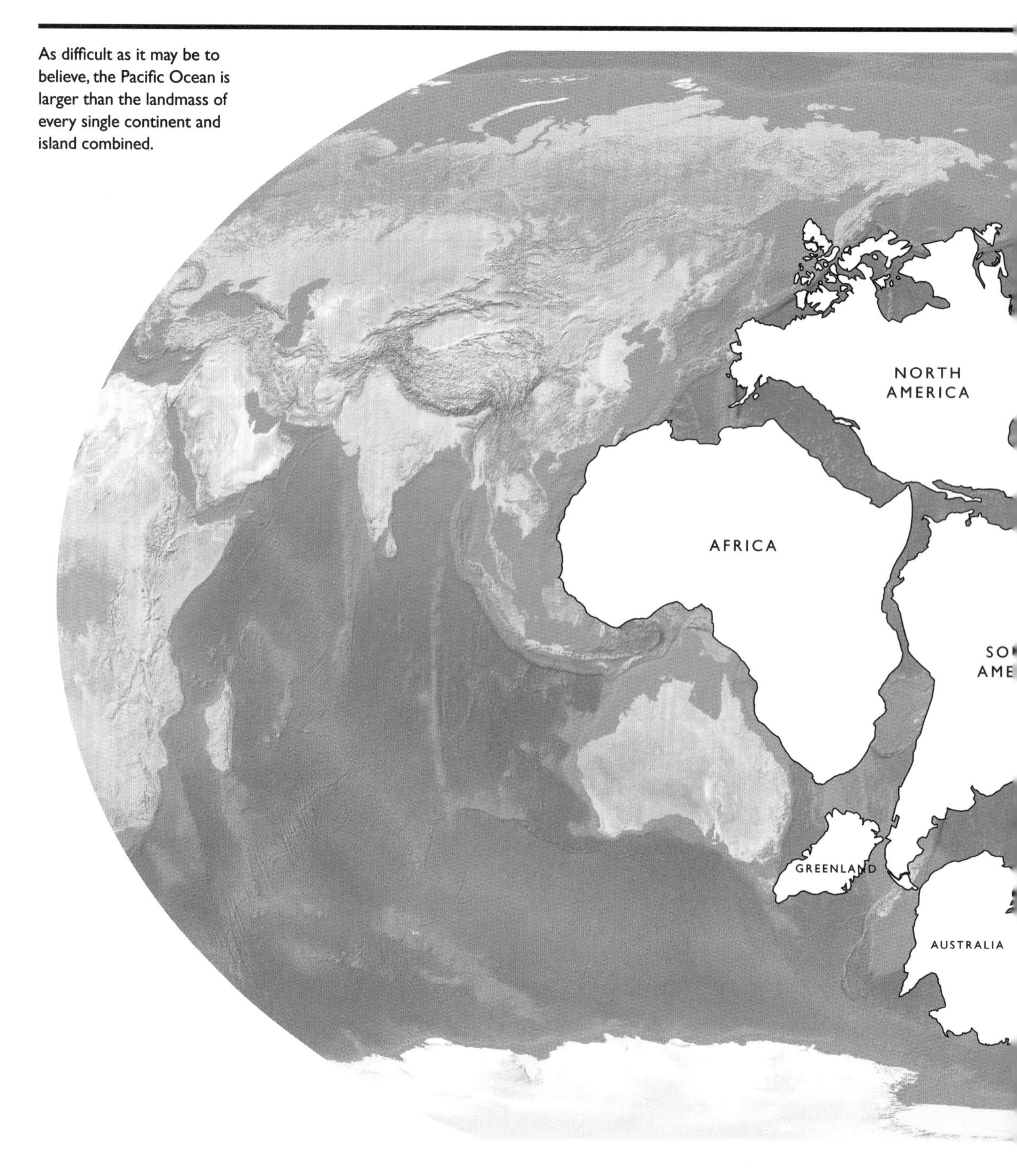

The Pacific Ocean covers 63.8 million square miles.

Earth's land mass covers 57.3 million square miles.

60 The **20 largest islands** in the world compared

Here are a few more facts about the islands below:

- Canada owns four of the top twenty islands, while Indonesia owns all or part of five.
- The least populated is Canada's Ellesmere Island, with just 191 people.
- The most populated is Indonesia's Java, with 145 million people.
- The US does not have a single island in the top twenty, though Hawai'i (Big Island) is the seventy-sixth largest in the world.
- Only three islands belong to more than one country—New Guinea, Borneo, and Ireland.
- Australia, three times the size of Greenland, is considered a continent rather than an island.

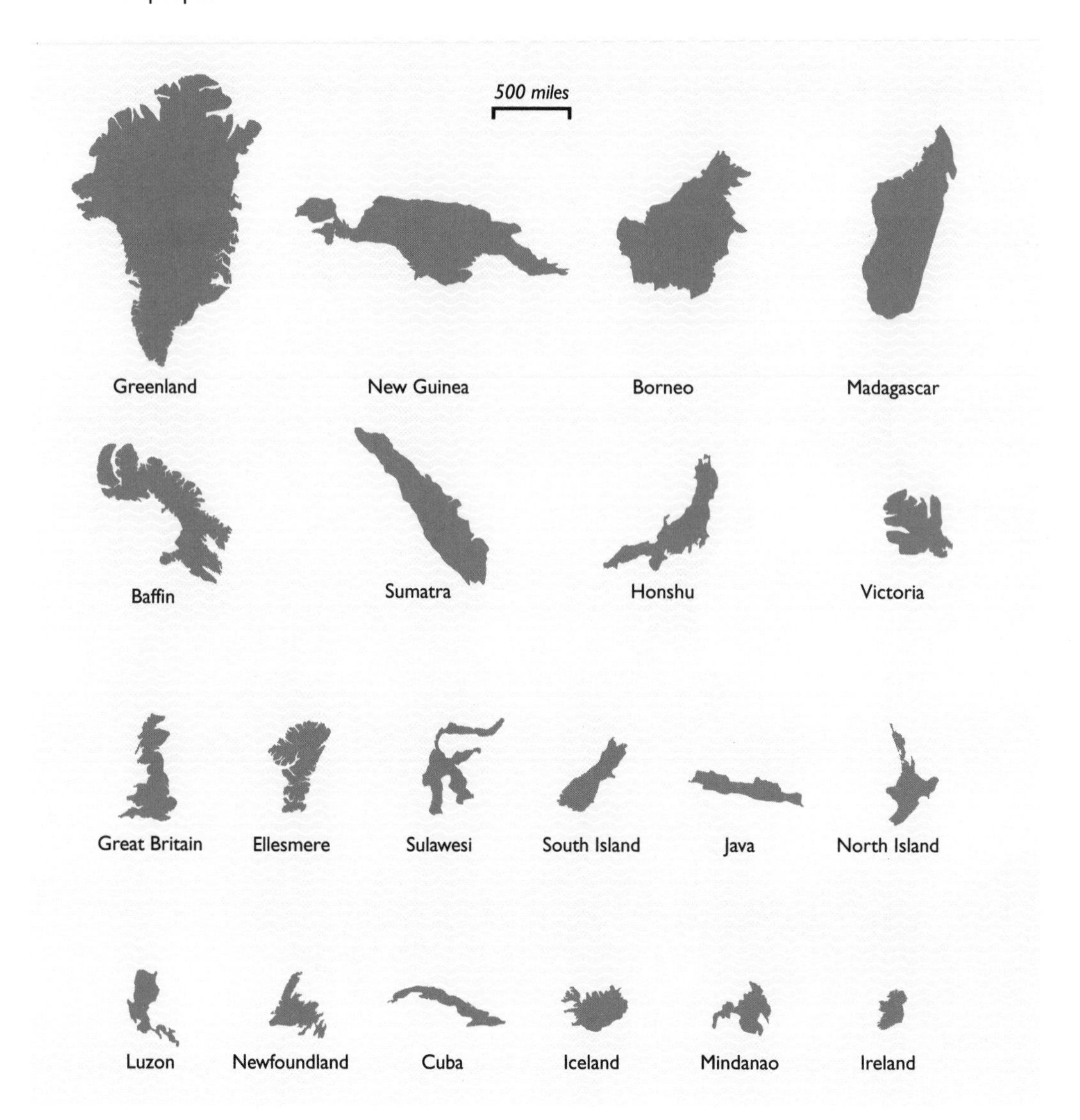

61 The **Pan-American Highway:** the longest road in the world

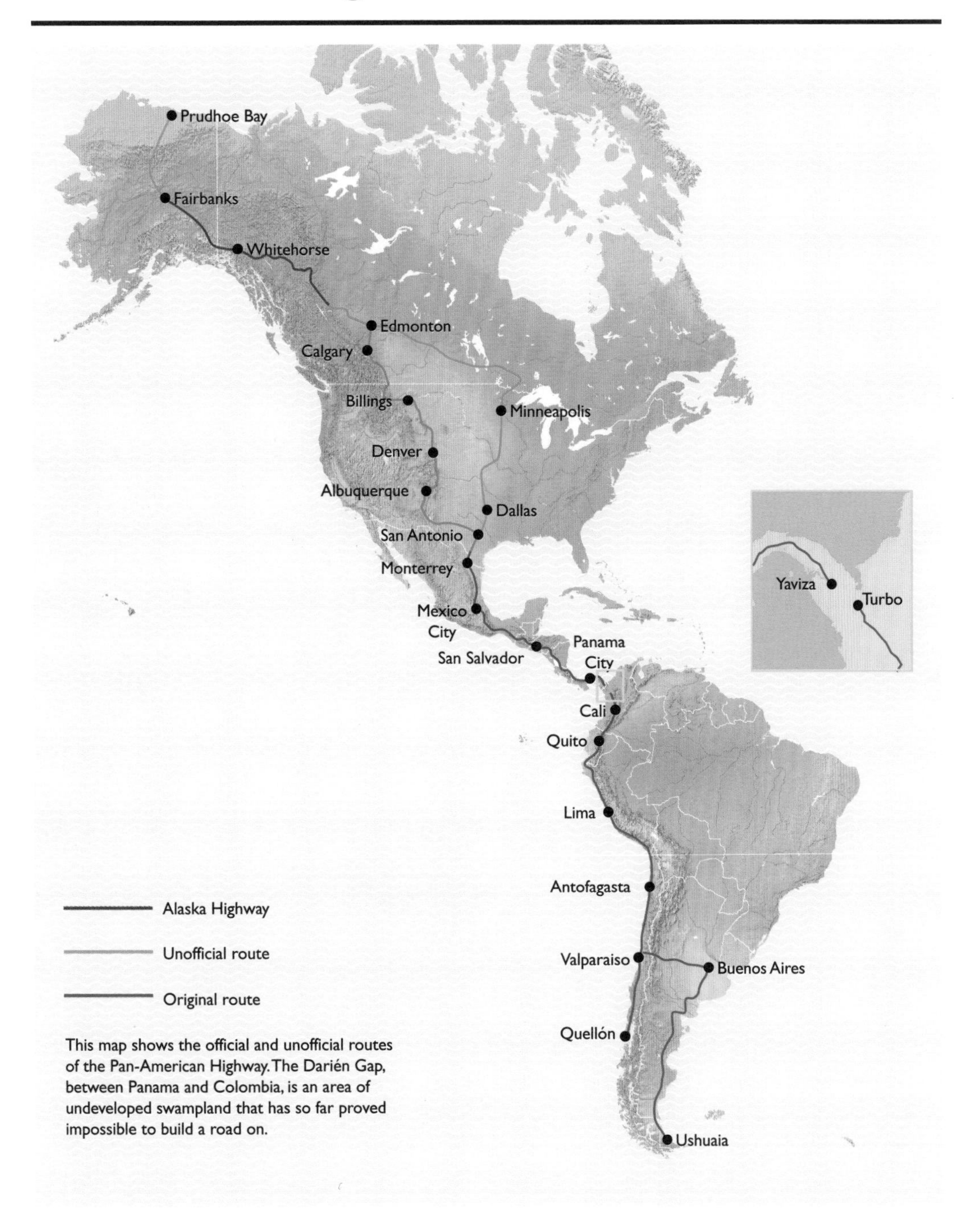

This map shows the official and unofficial routes of the Pan-American Highway. The Darién Gap, between Panama and Colombia, is an area of undeveloped swampland that has so far proved impossible to build a road on.

62 More people live **inside this circle** than outside of it

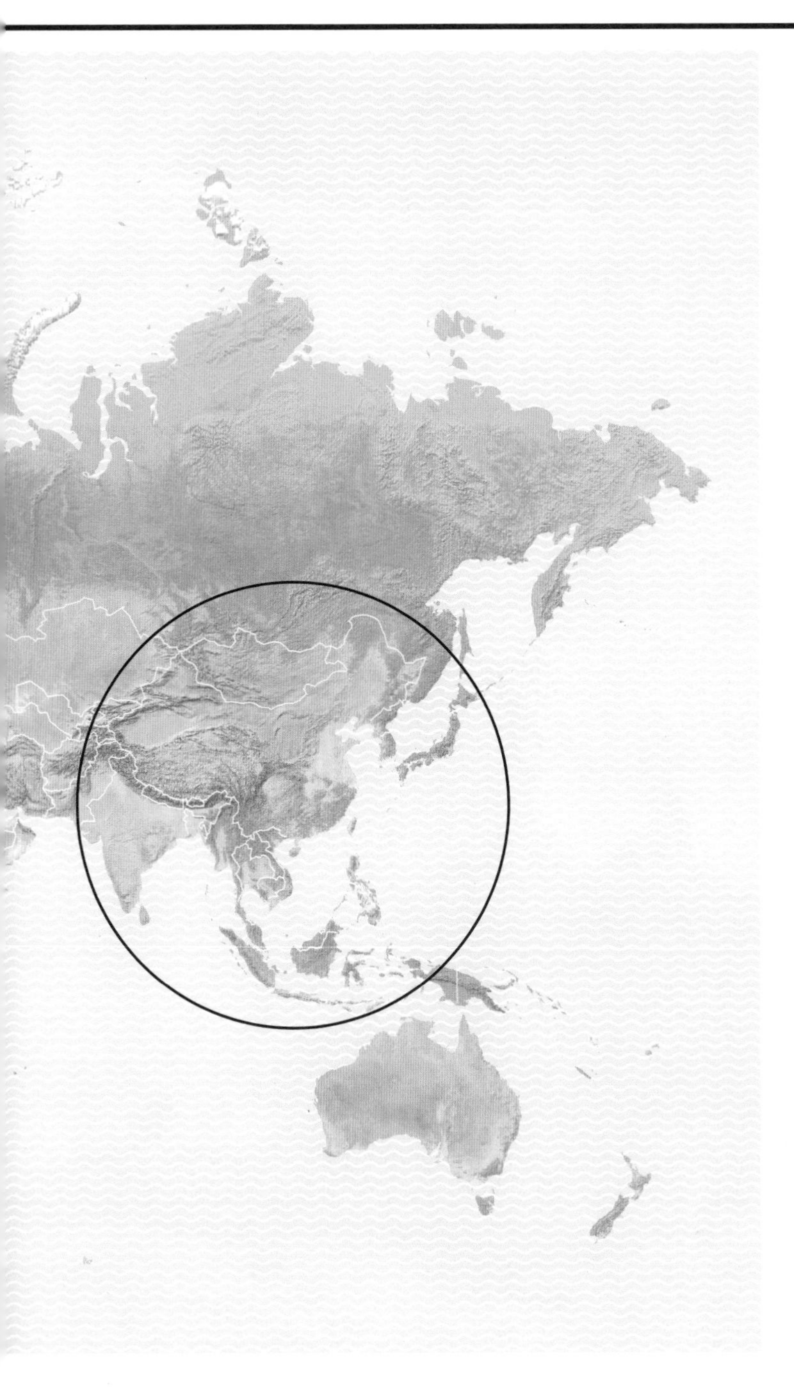

More Muslims inside than outside this circle

More Hindus inside than outside this circle

More Buddhists inside than outside this circle

More Communists inside than outside this circle

This circle contains one of the least densely populated countries in the world (Mongolia—5 people per mi^2)

This circle contains the highest mountain (Everest)

This circle contains the deepest ocean trench (Mariana)

63 **Antipodes** world map—or, why you can't dig to China from the US

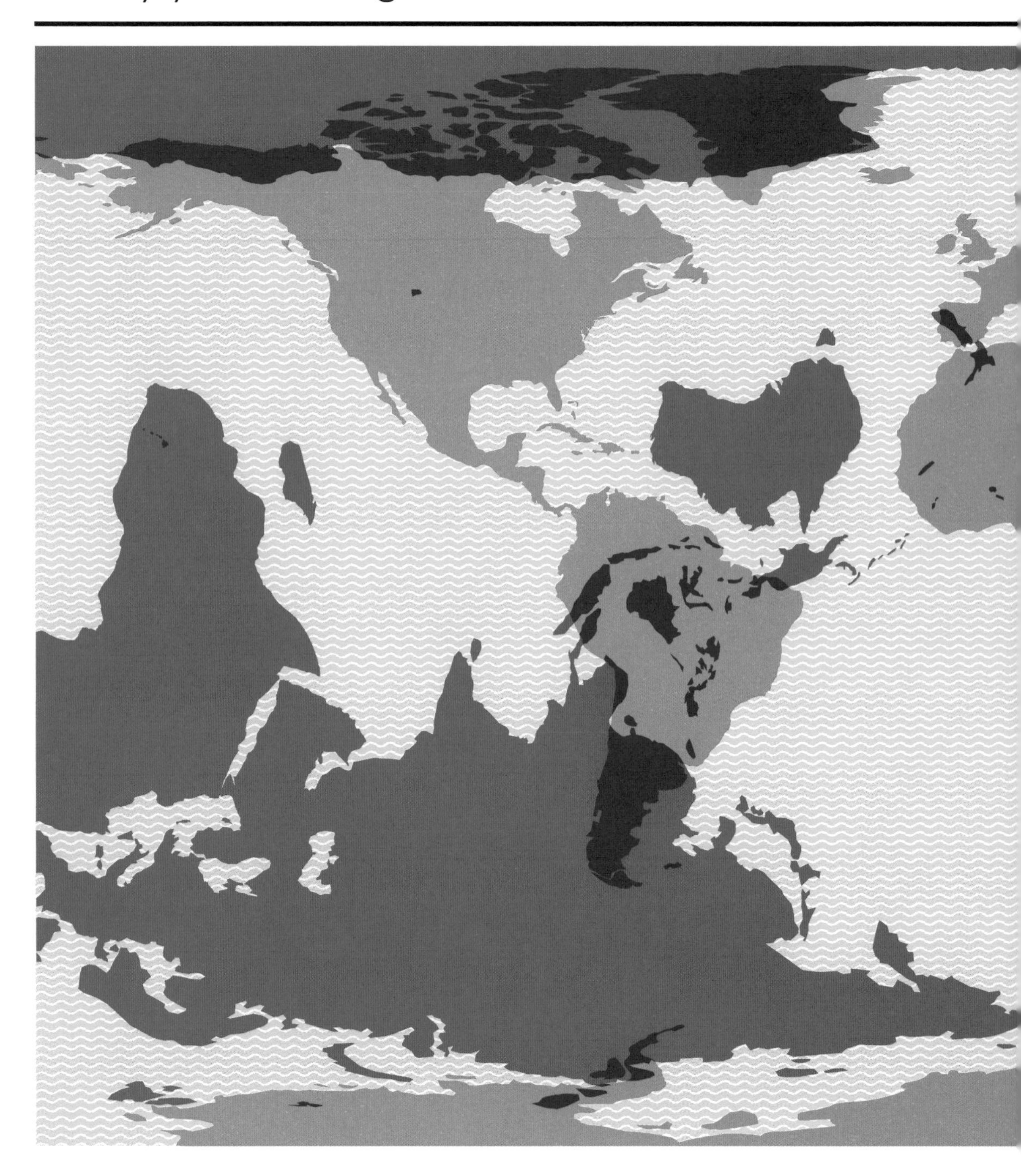

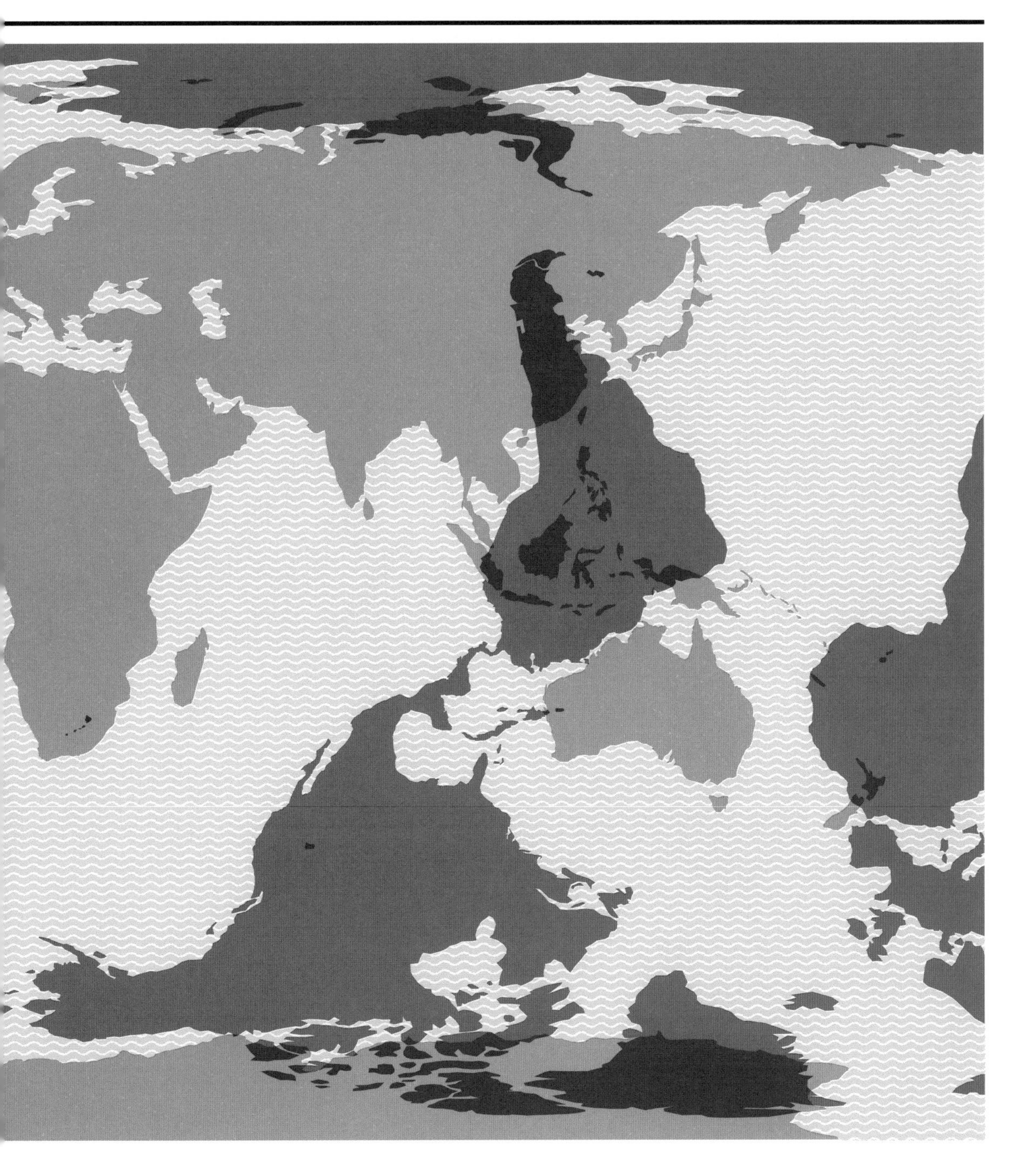

64 The world in a **mirror**

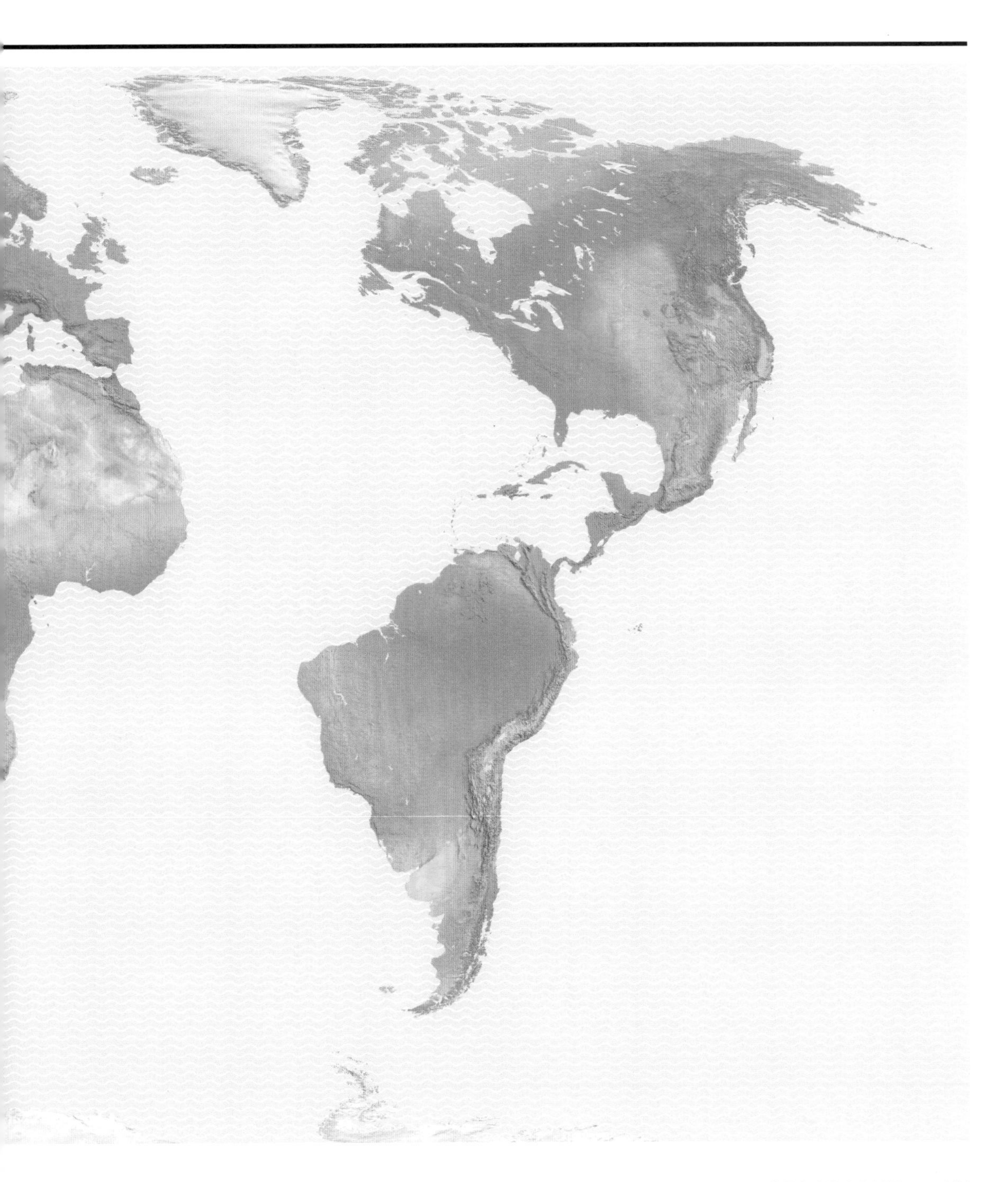

65 The world's **time zones**

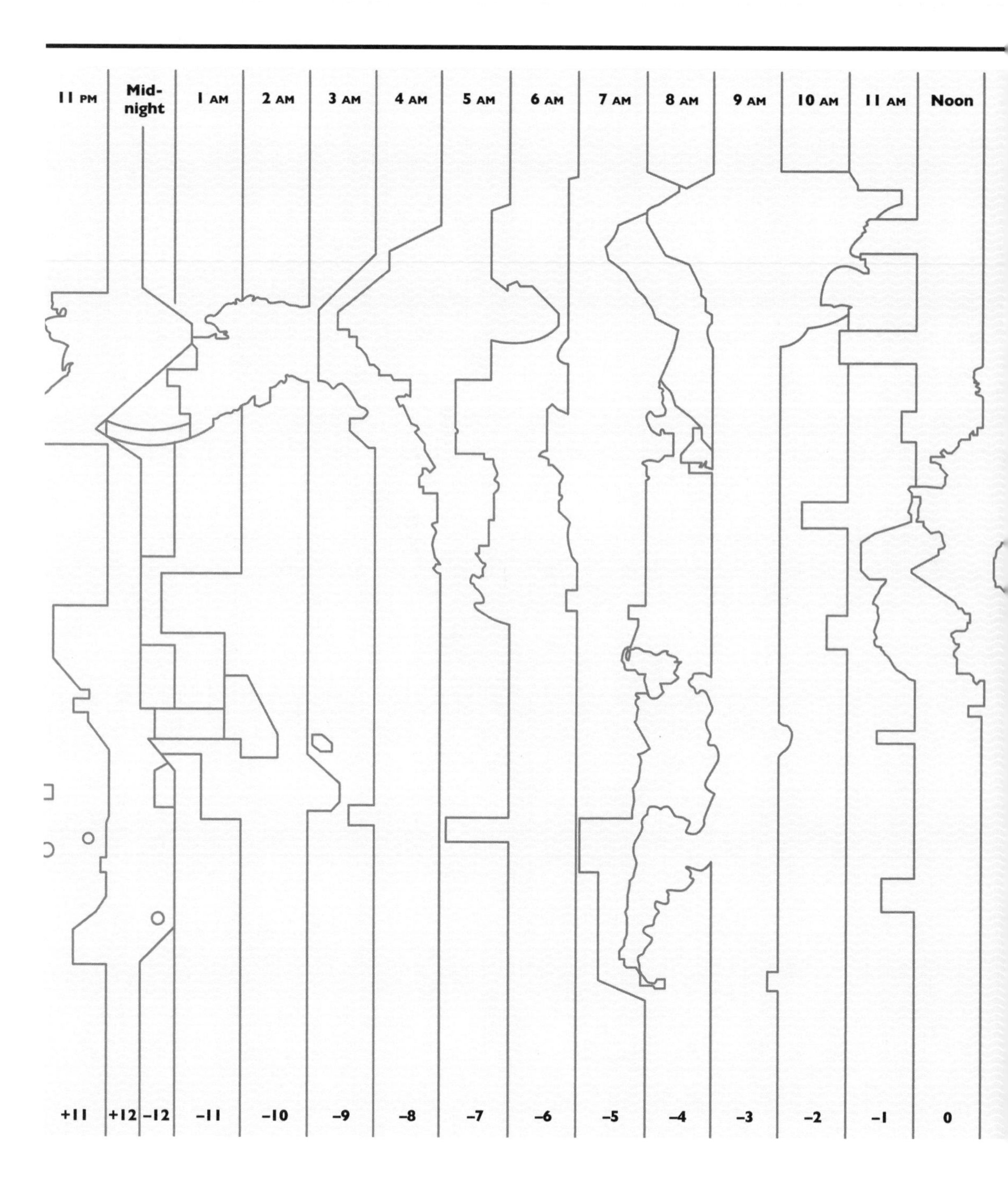

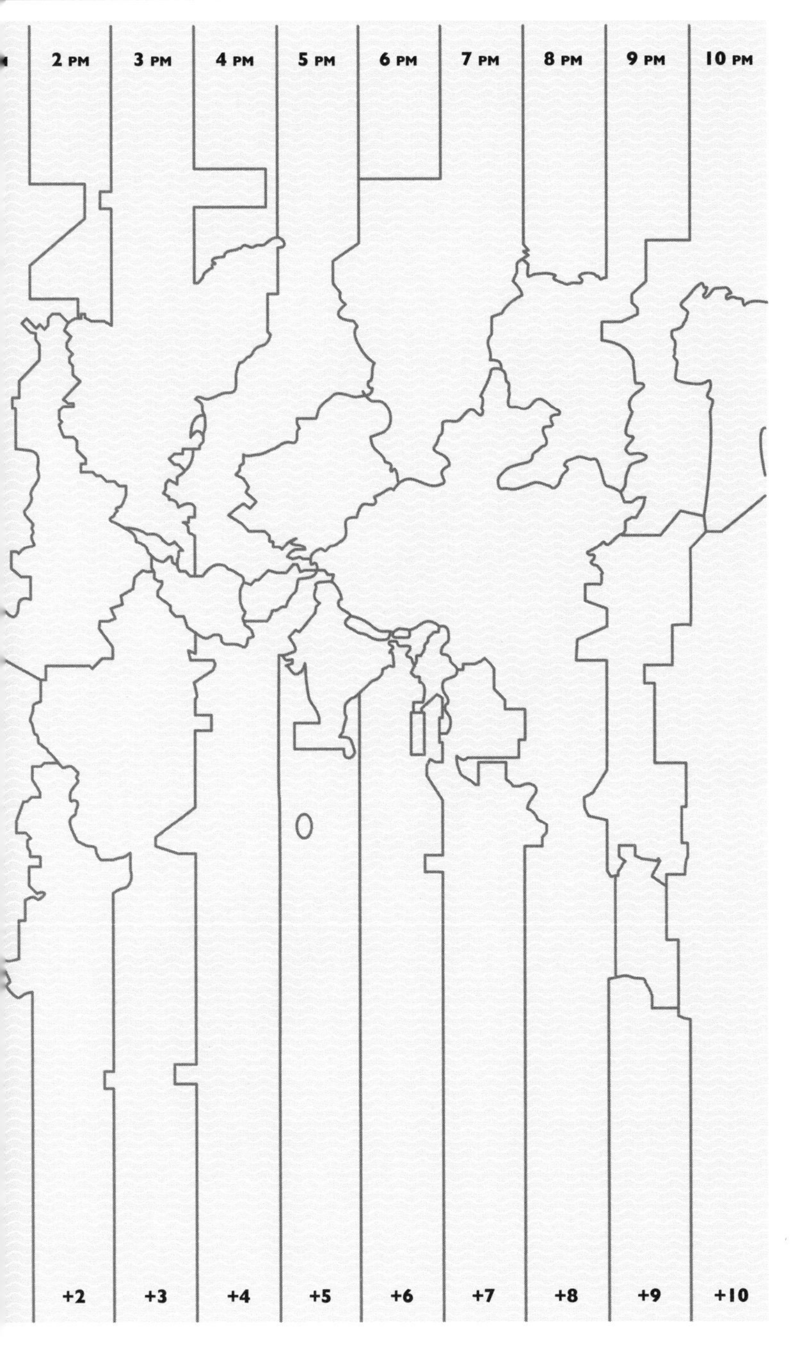

A few countries and territories have clear outlines on the map due to time zone decisions, including:

- China, which officially has just one time zone for the whole country (UTC+08:00).
- Argentina, which is on UTC-3:00 time (it should be UTC-4:00 or UTC-5:00) and which also only sometimes observes daylight savings time.
- Iceland, which is on GMT despite being significantly to the west of Greenwich.
- The Canadian province of Newfoundland, which has a time zone that is 3:30 mins behind UTC (UTC-3:30).
- India and Sri Lanka, which both use Indian Standard Time (UTC+5:30).
- Iran, which uses Iran Standard Time (UTC+03:30).
- Nepal, which uses Nepal Standard Time (UTC+5:45), which means if it's 12 PM UTC then it will be 5:45 PM in Nepal.

66 The world's five longest **domestic nonstop flights**

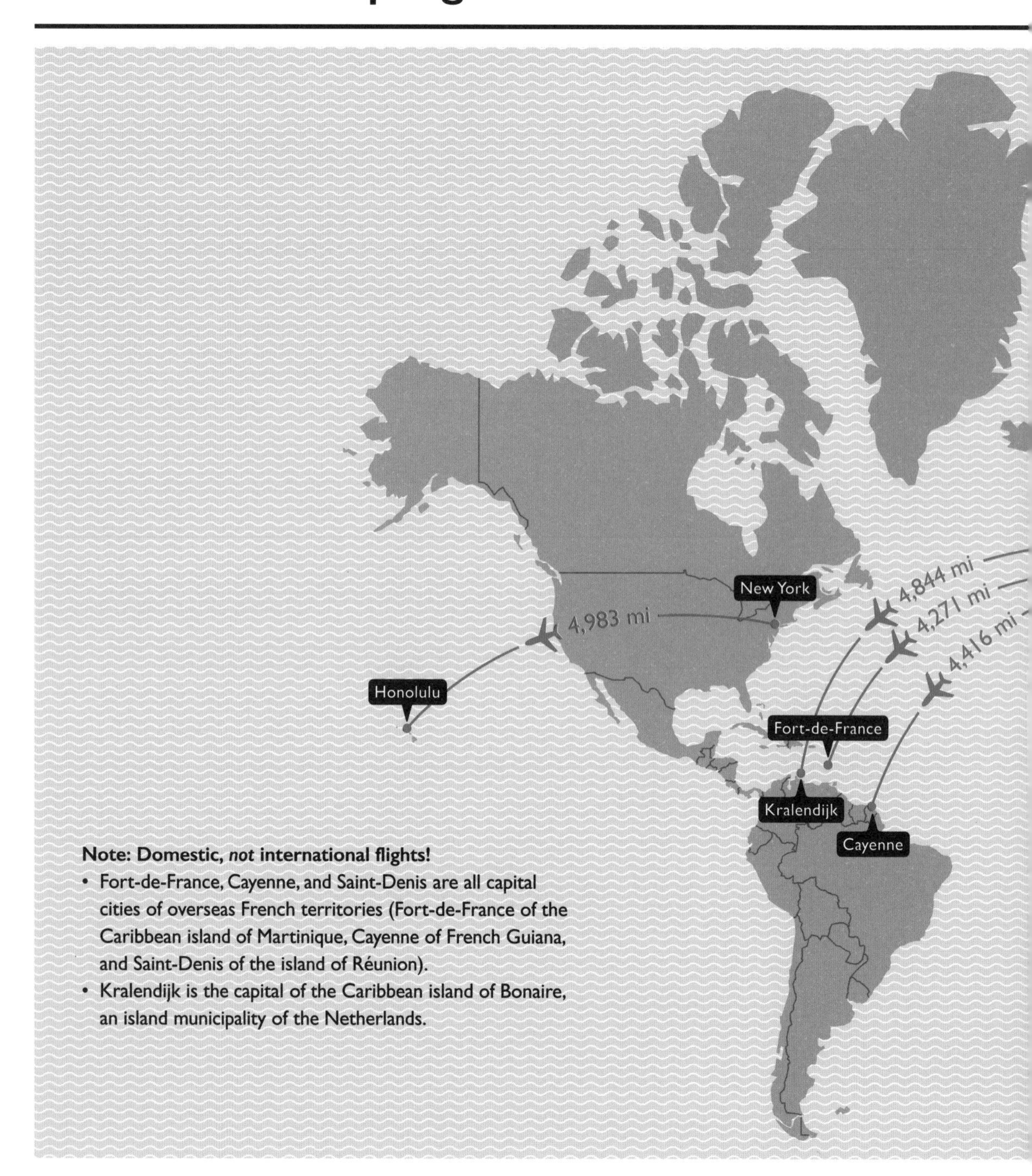

Note: Domestic, *not* international flights!

- Fort-de-France, Cayenne, and Saint-Denis are all capital cities of overseas French territories (Fort-de-France of the Caribbean island of Martinique, Cayenne of French Guiana, and Saint-Denis of the island of Réunion).
- Kralendijk is the capital of the Caribbean island of Bonaire, an island municipality of the Netherlands.

Amsterdam
Paris
5,809 mi
Saint-Denis

67 **Travel times** from London in 1914

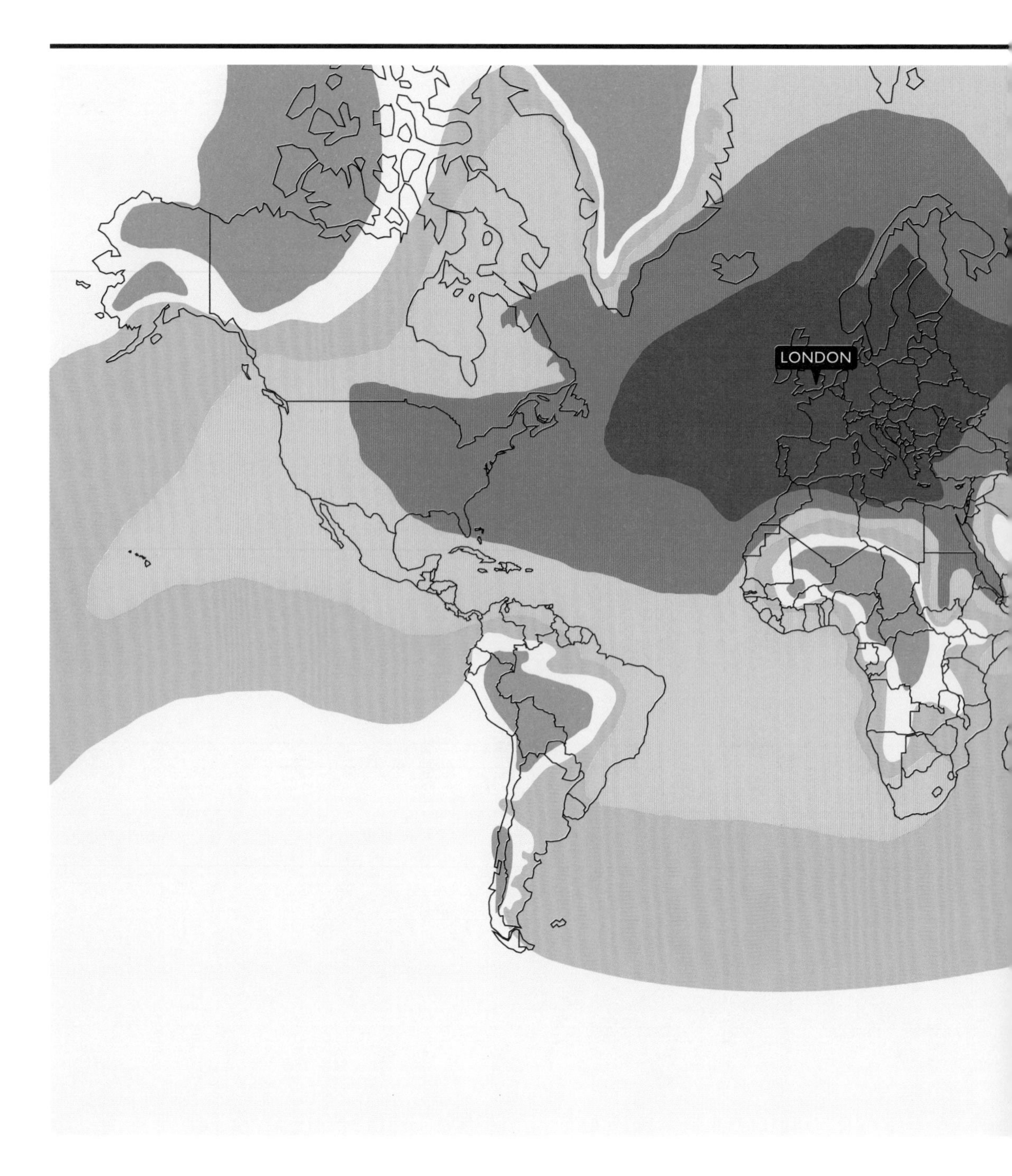

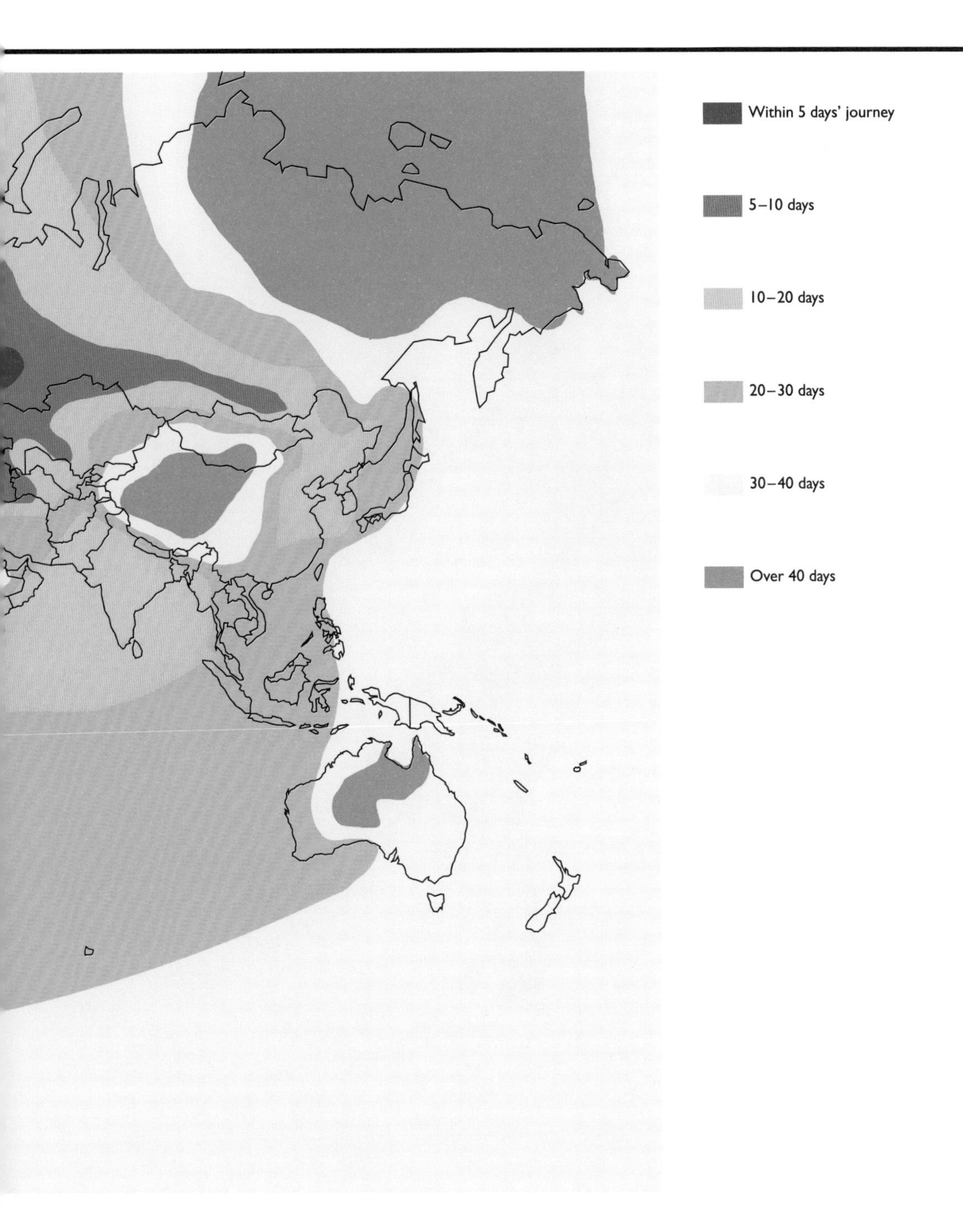
Within 5 days' journey
5–10 days
10–20 days
20–30 days
30–40 days
Over 40 days

68 **Travel times** from London in 2016

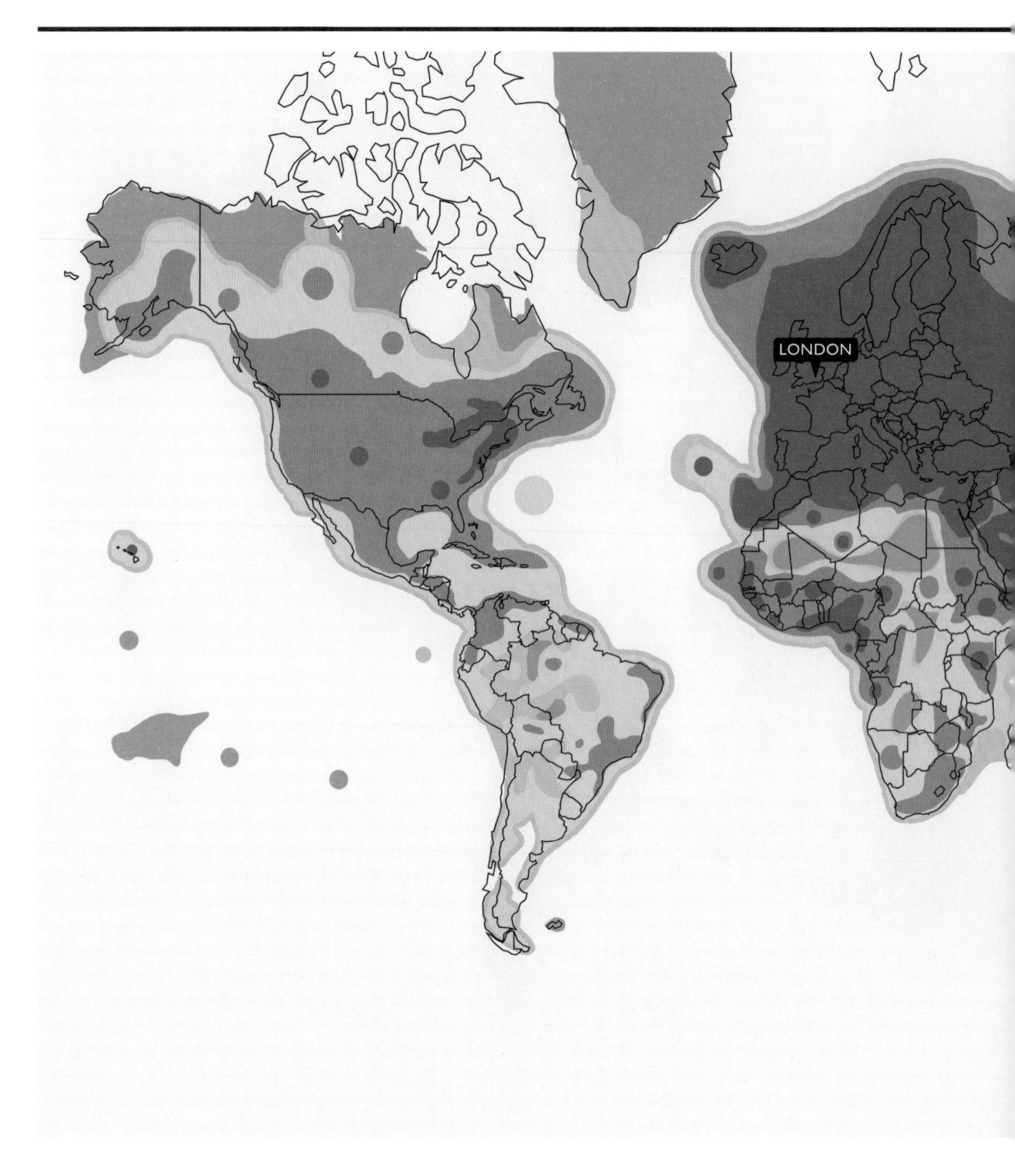

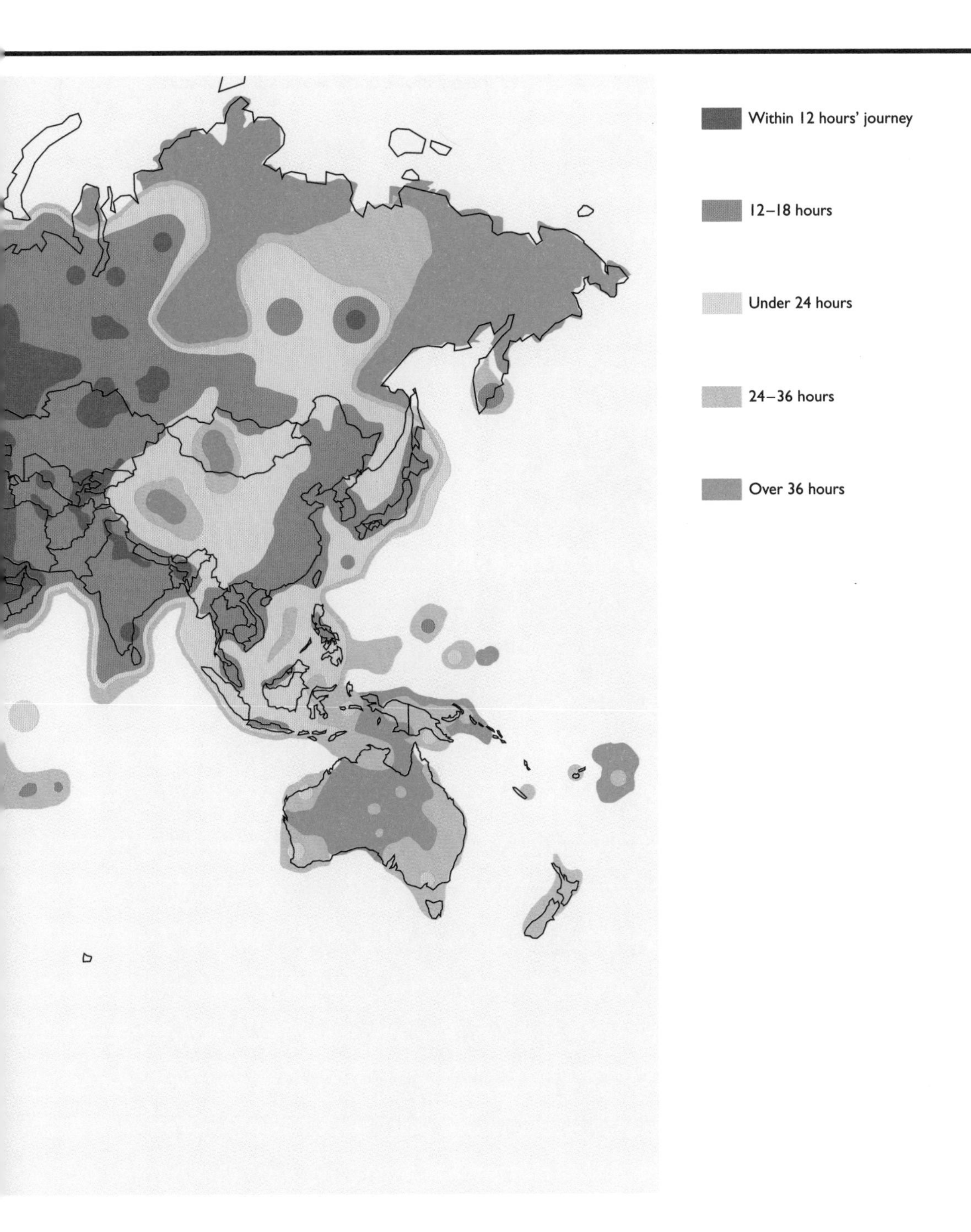
Within 12 hours' journey
12–18 hours
Under 24 hours
24–36 hours
Over 36 hours

69 **Luxembourg** is not a microstate!

Luxembourg is 998.6 square miles with a population of 576,249, making it one of the world's smallest states; 168th by size or 164th by population.

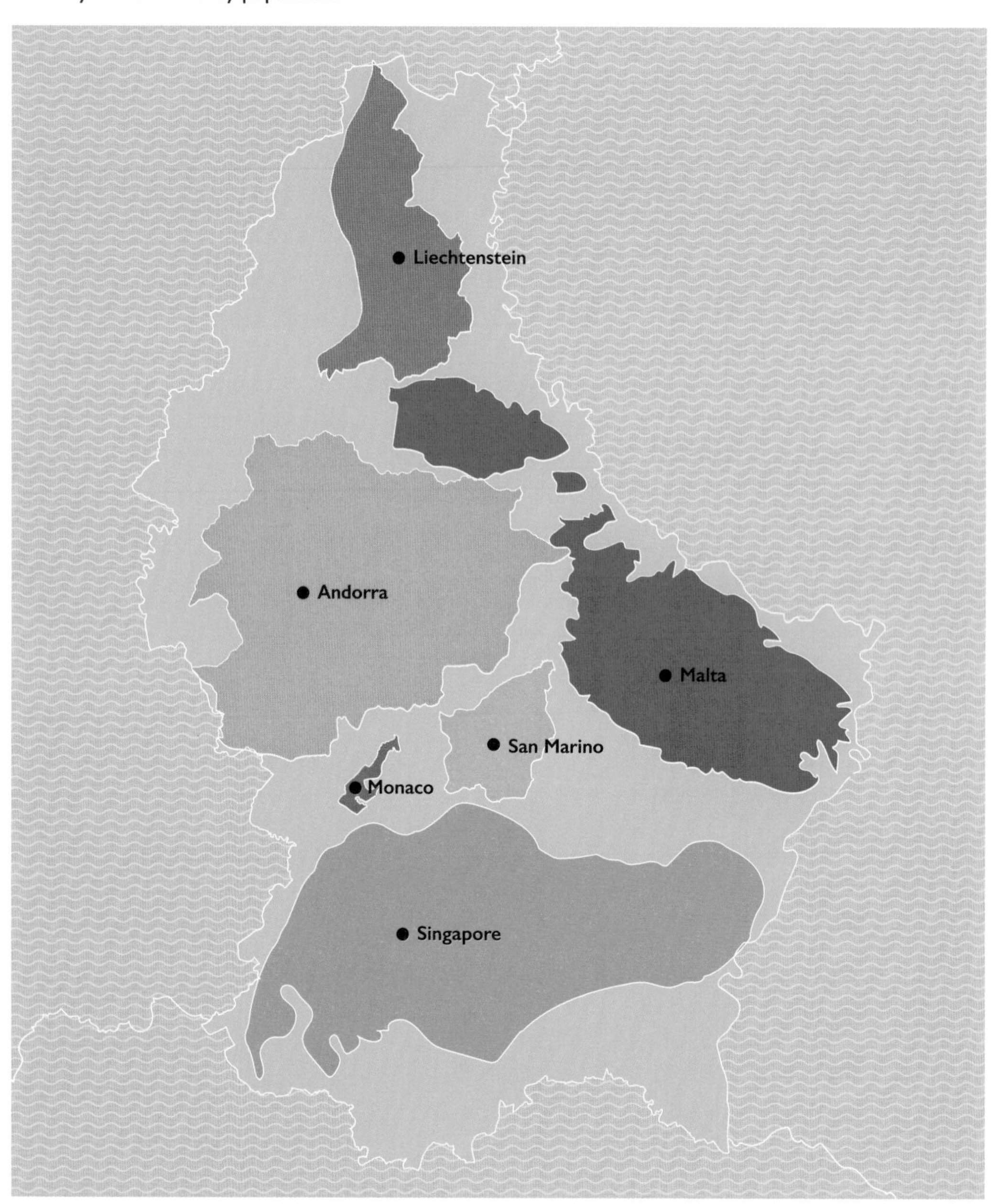

70 All roads lead to **Rome***

* The saying still holds true across the European continent, where all major arteries connect to freeways that lead to Rome.

6

HISTORY

71 European map of the **unexplored world** (1881)

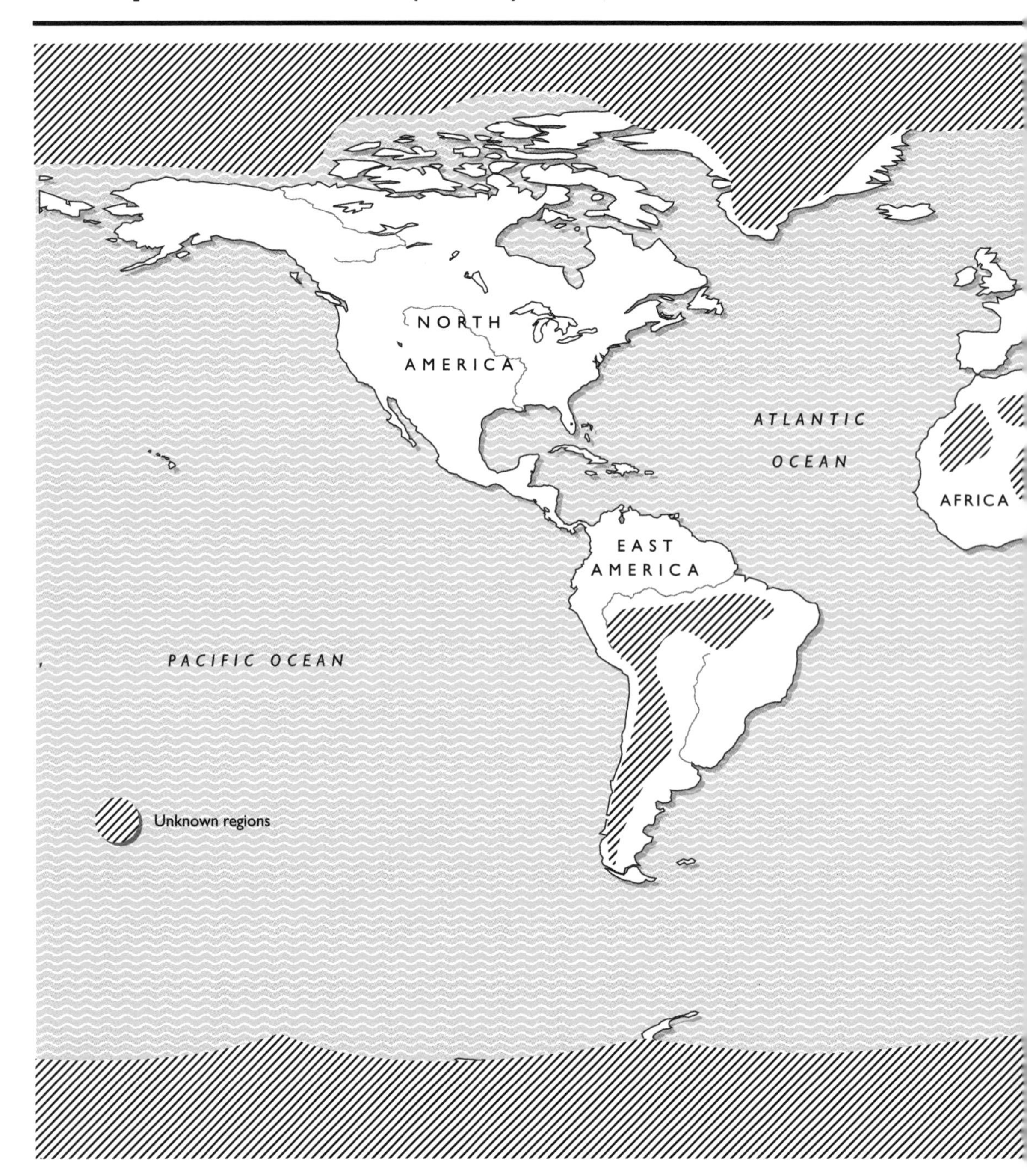

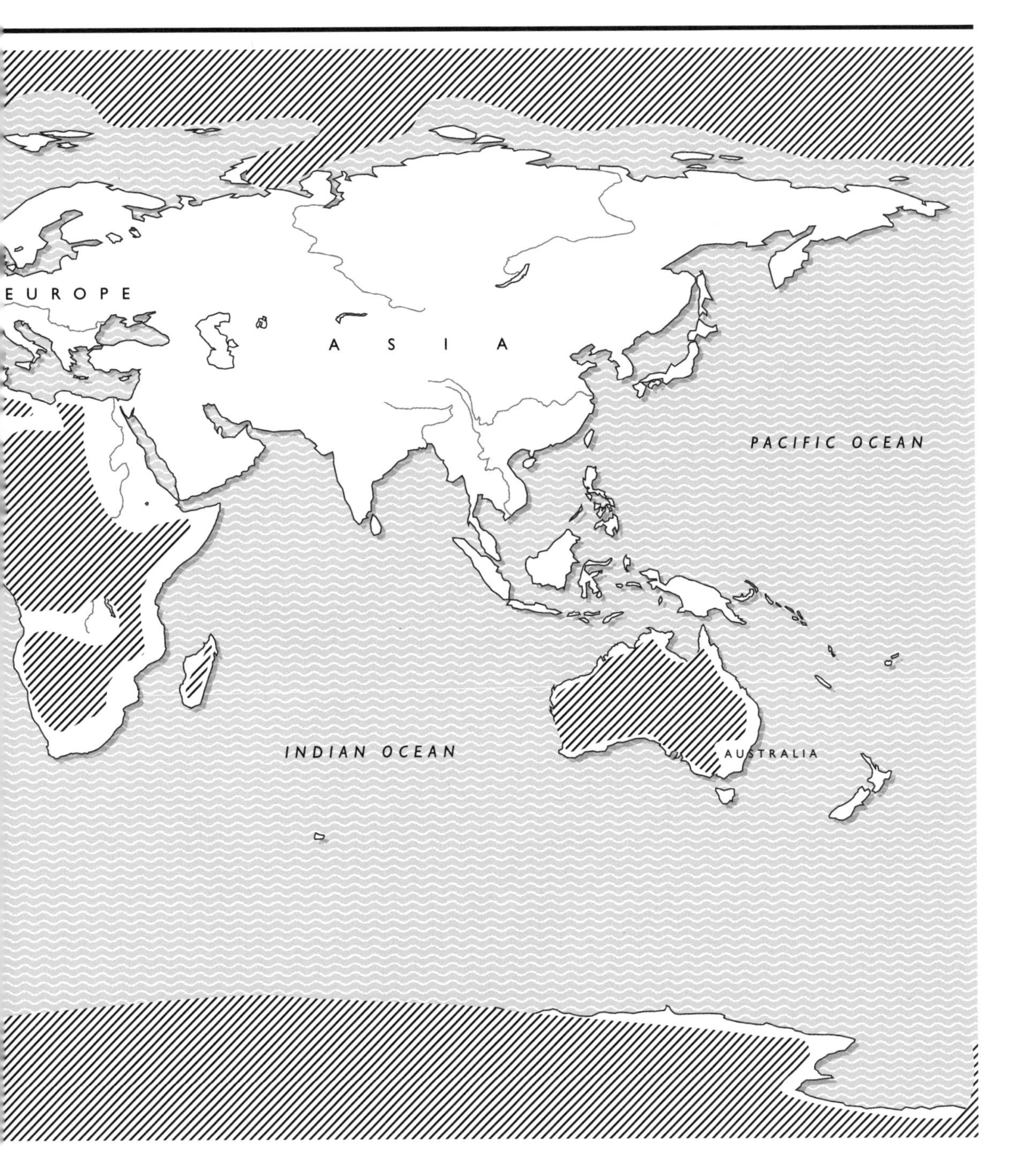
EUROPE
ASIA
PACIFIC OCEAN
INDIAN OCEAN
AUSTRALIA

72 **Colonial Africa** on the eve of World War I

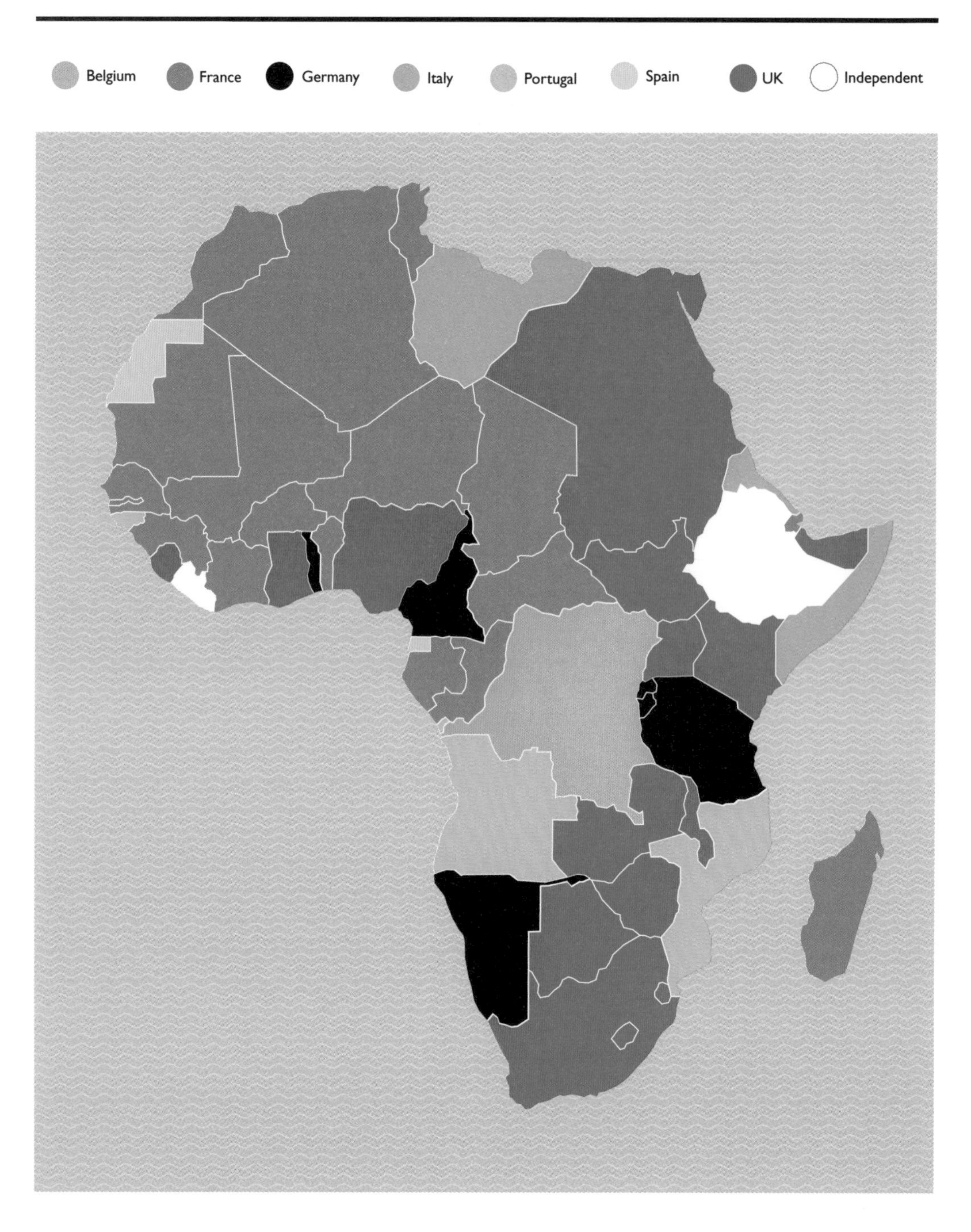

73 Map of the **entire internet** in December 1969

UCLA Network Measurement Center at the University of California, Los Angeles

SRI NLS system at SRI International in Menlo Park, California

UCSB Culler-Fried Interactive Mathematics Center at the University of California, Santa Barbara

UTAH University of Utah School of Computing

74 If the **Roman Empire reunited—** using modern borders

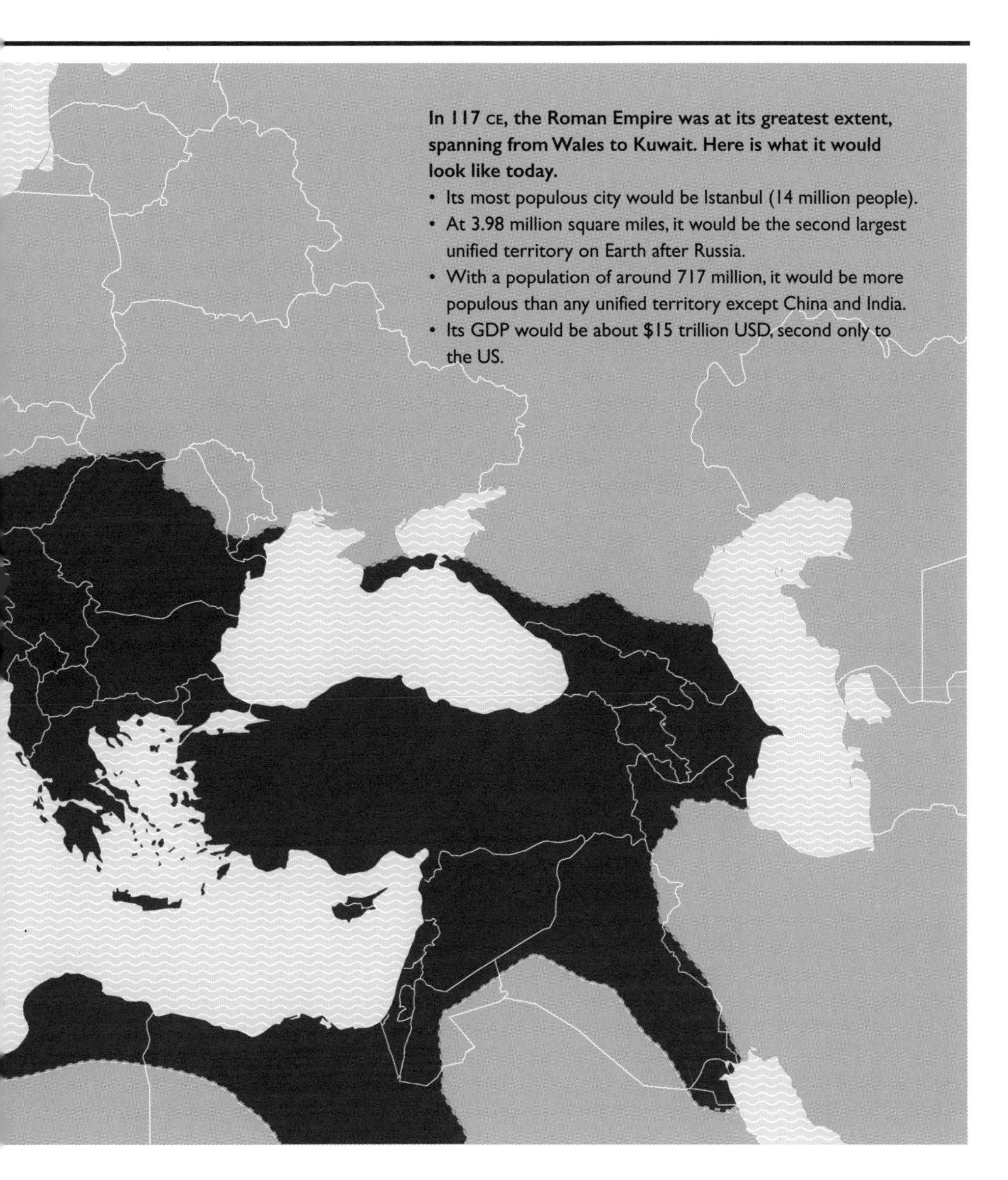

In 117 CE, the Roman Empire was at its greatest extent, spanning from Wales to Kuwait. Here is what it would look like today.

- Its most populous city would be Istanbul (14 million people).
- At 3.98 million square miles, it would be the second largest unified territory on Earth after Russia.
- With a population of around 717 million, it would be more populous than any unified territory except China and India.
- Its GDP would be about $15 trillion USD, second only to the US.

75 The first proposed **map*** of **Pakistan** and **the partition of India**

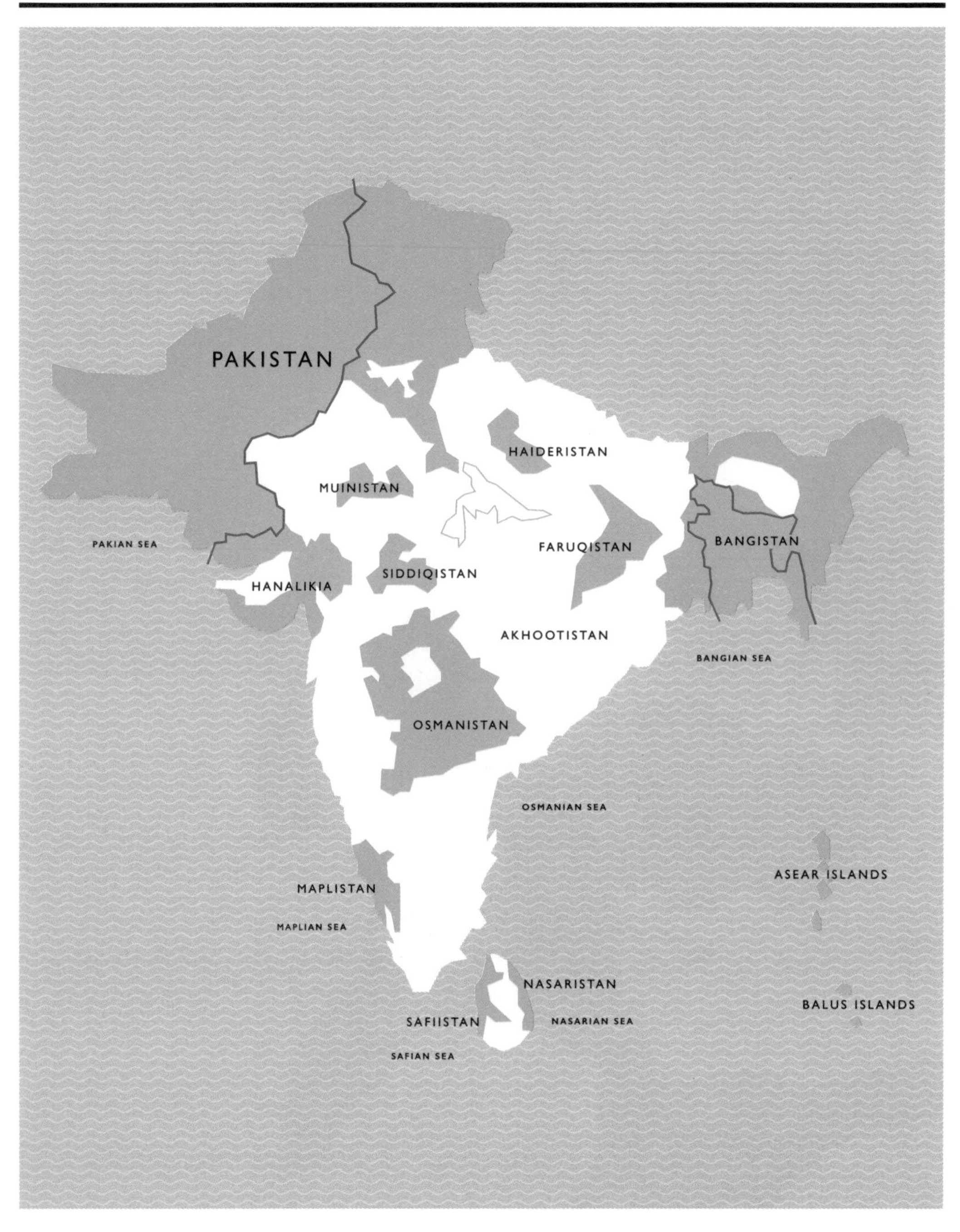

* By Choudhry Rahmat Ali in his 1935 book *Pakistan: The Fatherland of the Pak Nation*

76 If the **Mongol Empire reunited**

The Mongol Empire was the largest contiguous land-based empire in history. Here is what the empire in 1279 would look like on today's globe.

- It would be the world's largest country at 13.5 million square miles (twice the size of Russia).
- It would be the world's most populous country, with 2.2 billion people (60 percent more than China).
- It would be the world's largest economy at $29.2 trillion USD (accounting for purchasing power parity; 70 percent bigger than either the US or China).
- It would have the world's largest army with 9.75 million active military personnel (6.5 times larger than the military of the US).
- Its biggest city would be Shanghai, which, while not the largest metropolitan area on Earth, is the largest city based on city limits.

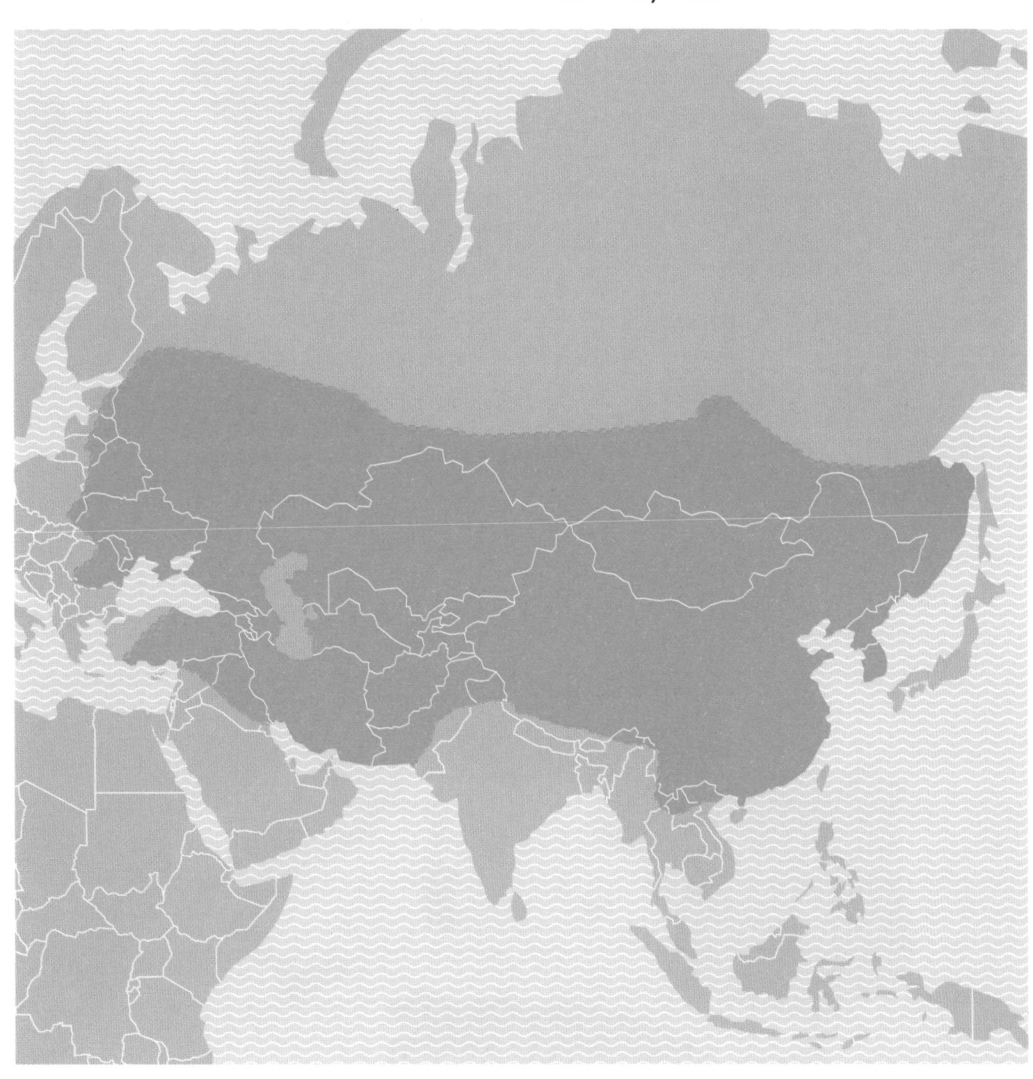

77 When **Great Britian** was connected to **continental Europe**

Around 18,000 years ago, the land mass known as Doggerland connected Britain with continental Europe. Rising sea levels flooded Doggerland in 6,500 to 6,200 BCE.

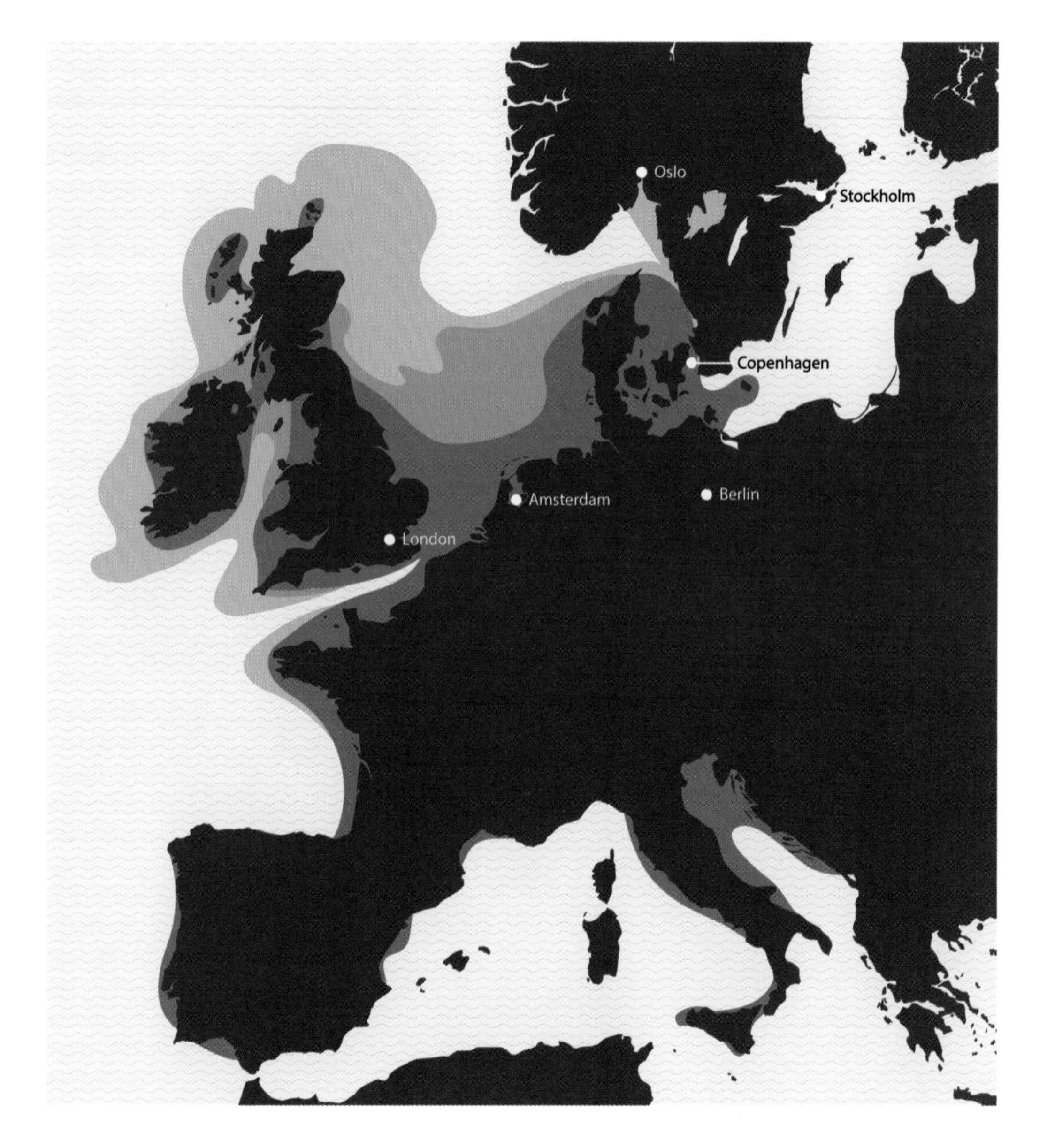

78 Where were the **Seven Wonders** of the Ancient World?

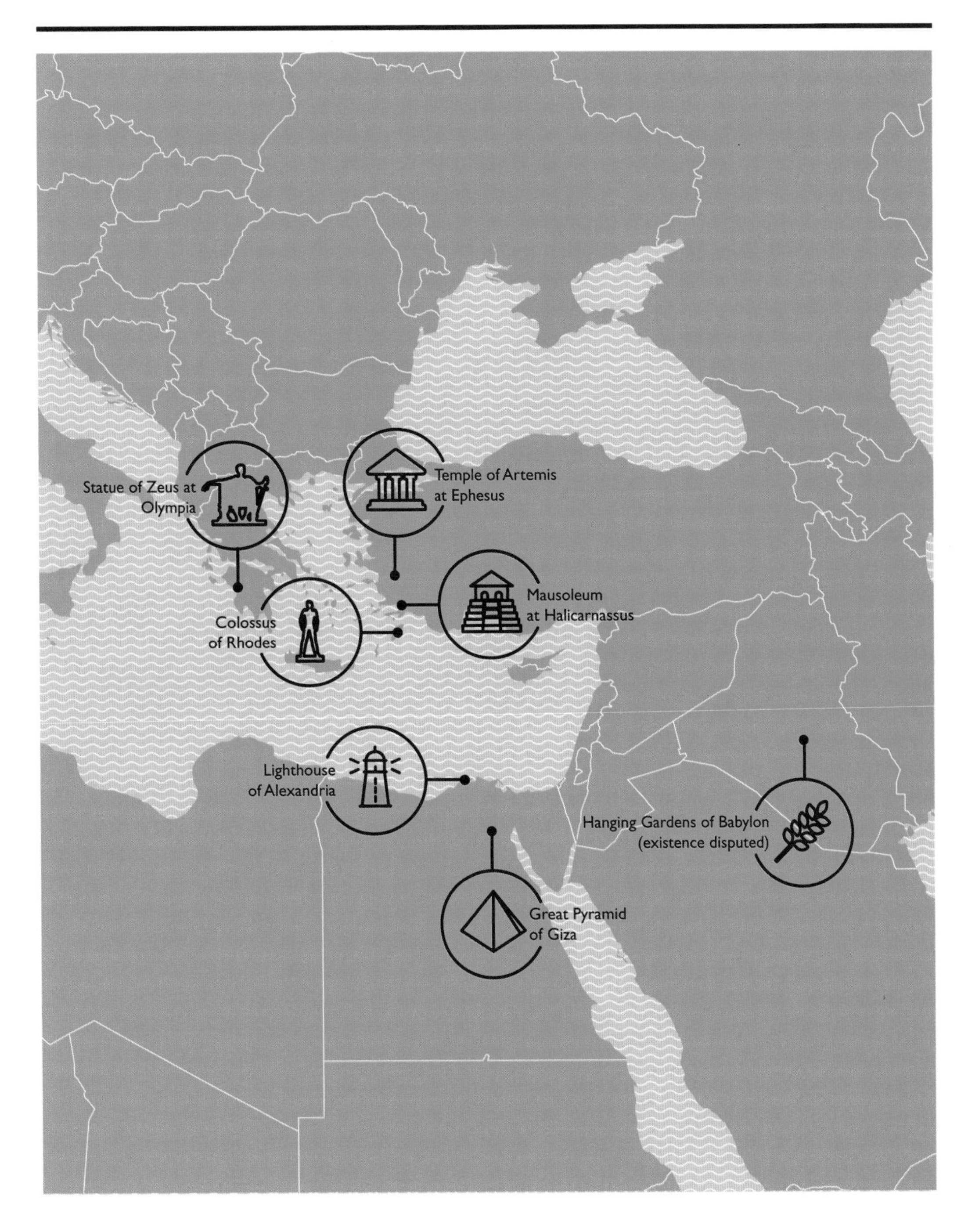

79 **WW1 casualties** as a percentage of prewar populations

Includes both military and civilian casualties

- No data
- 0–1%
- 1–2%
- 2–4%
- 4–10%
- >10%

80 **WWII casualties** as a percentage of prewar populations

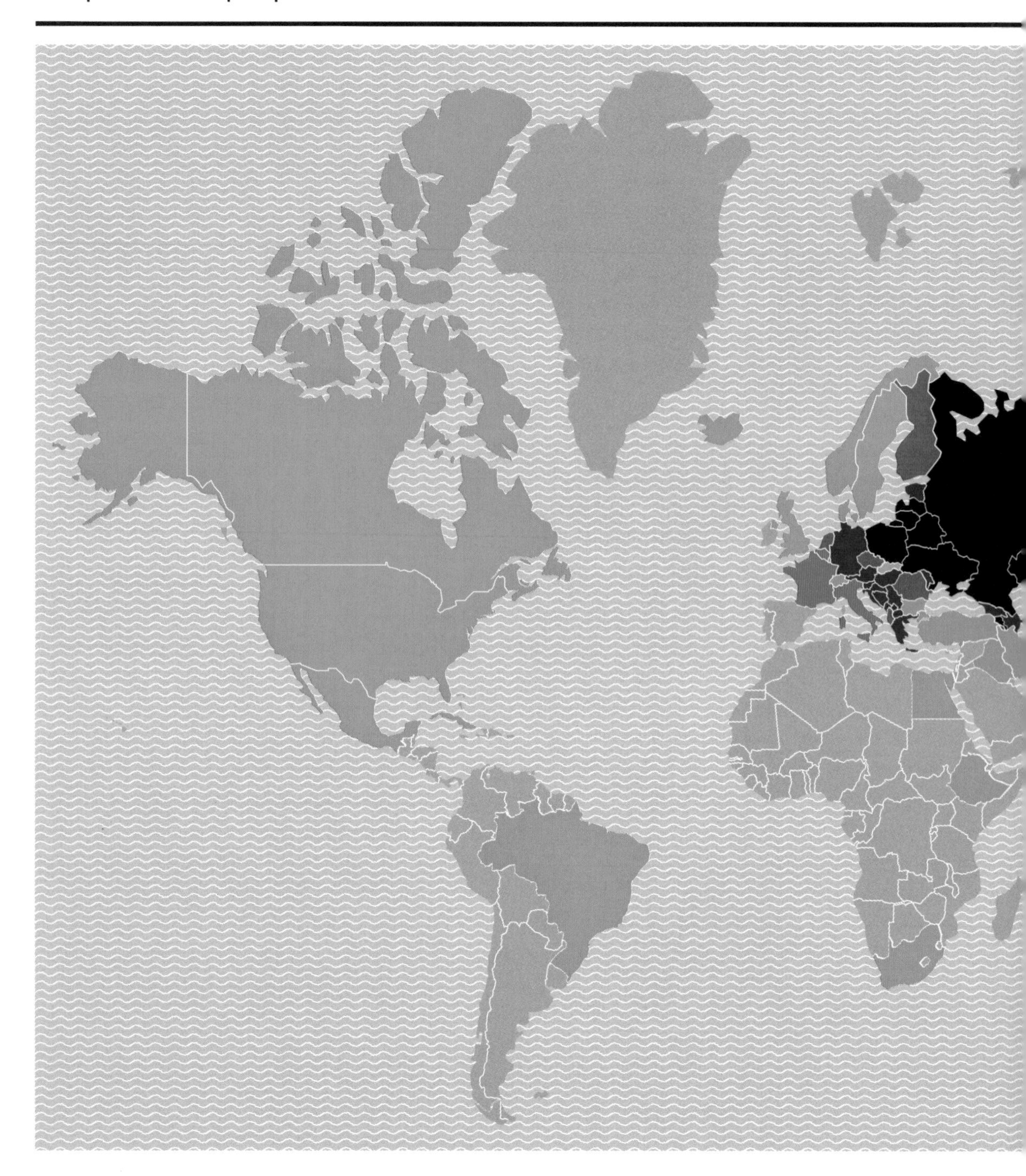

Includes both military and civilian casualties

- No data
- 0–1%
- 1–2%
- 2–4%
- 4–10%
- >10%

81 Countries that lost citizens on **9/11**

Countries with citizens who were victims of 9/11

Countries with citizens who were terrorists on 9/11

Countries with both citizens who were victims and citizens who were terrorists (Lebanon)

7

NATIONAL IDENTITY

82 Colors of **passports** around the world

Red

Green

Blue

Black

Britain's passport color is in the process of changing for purely political reasons.

83 Countries whose **flags** contain red and/or blue

Flag has
red and/or blue

Flag has
no red or blue

84 Flags of the world

MAYOTTE

85 If **European borders** were drawn by **DNA** instead of ethnicity

Using Y-chromosomal DNA (without mitochondrial DNA), this map shows the dominant haplogroup (group of genes inherited from a single parent) across Europe.

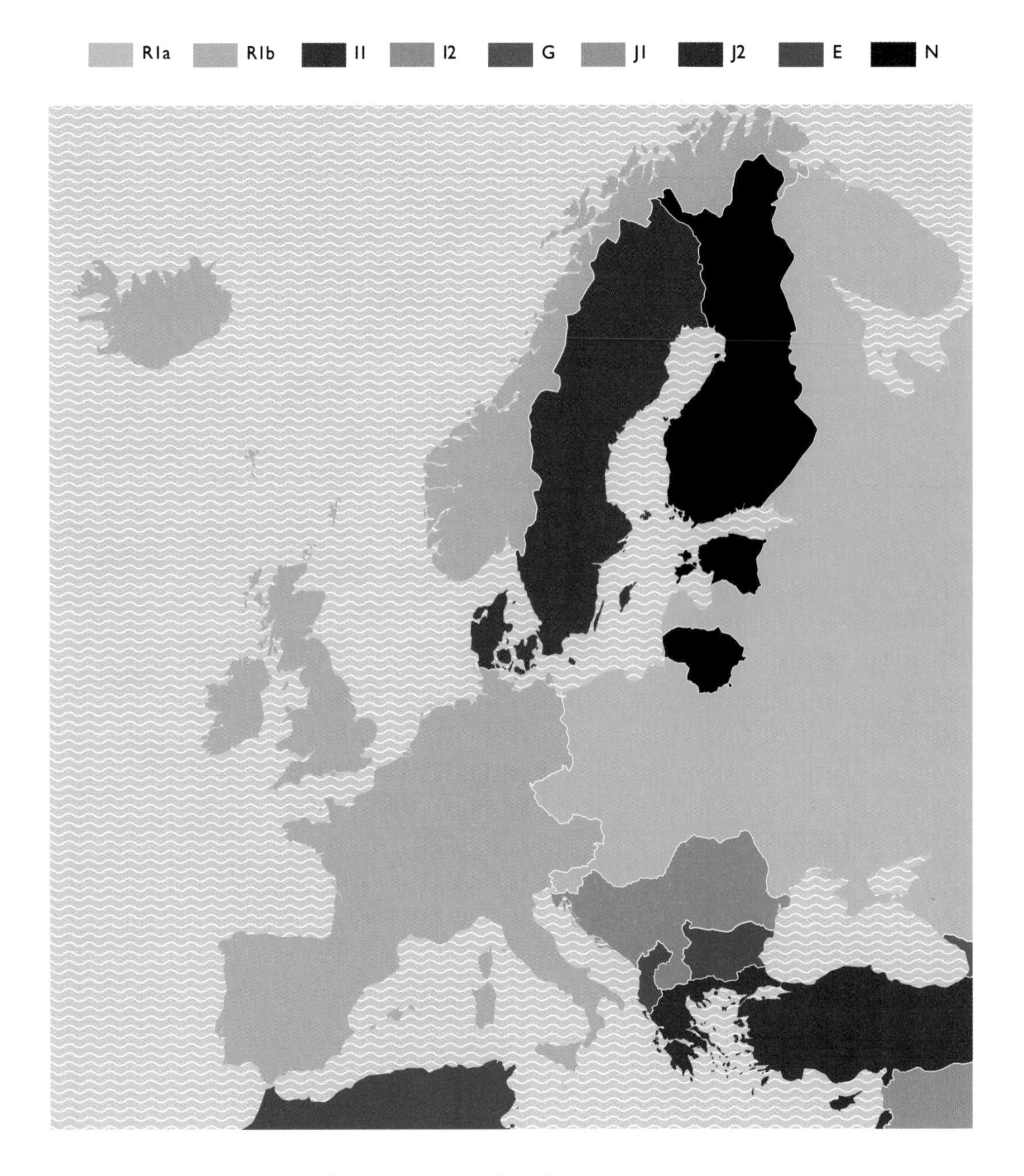

86 "Indian" isn't a language

India recognizes twenty-two regional languages. Missing from the key are Dogri and Sindhi, which are not the official state language or co-language of any one of India's twenty-nine states and union territories.

Arunachal Pradesh, Meghalaya, Nagaland, and Sikkim, all shaded in grey, list English as their state language but use a wealth of other languages and dialects. Mizoram also lists Mizo as a state language, but Mizo is not recognized as an official regional language by the Eighth Schedule to the Constitution of India.

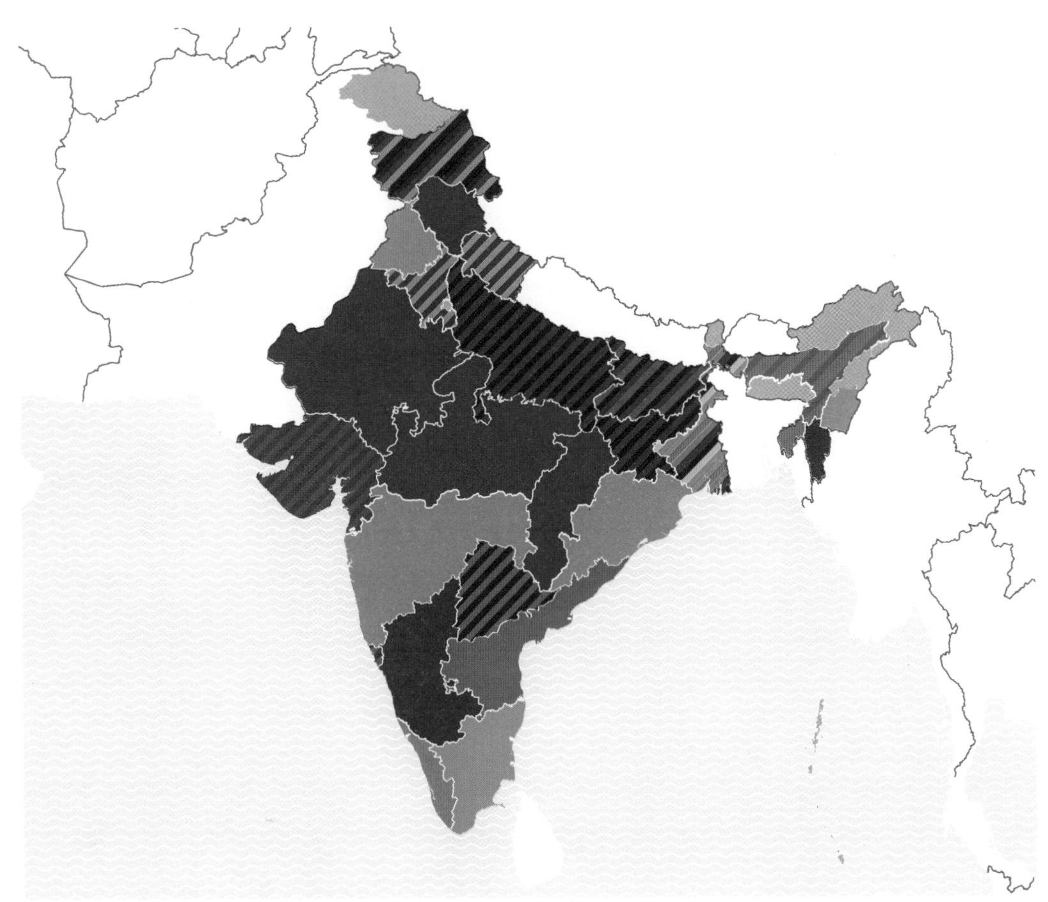

87 What do we mean when we say **"Asian"**?

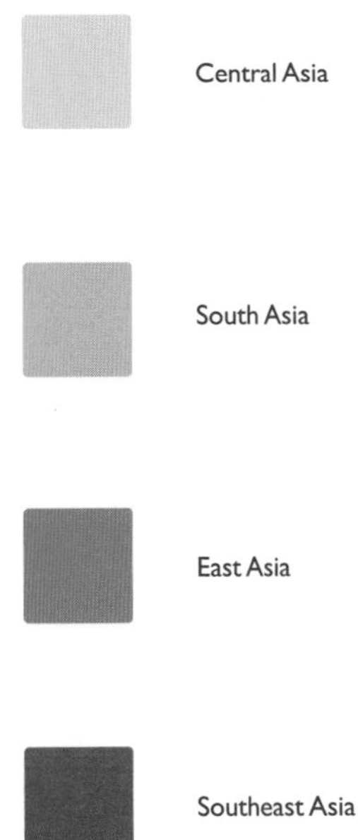

What about North and West? West Asia significantly overlaps with the Middle East, and North Asia comprises the Russian regions Siberia, Ural, and the Russian Far East, so the terms are rarely used.

8

CRIME AND PUNISHMENT

88 The US has as many **murders** annually as all the countries in blue combined*

* In 2017 (or 2016, when 2017 data was unavailable)

89 Homicide rates: Europe* vs. the US

Rate per 100,000 people

<1 | 1–2.99 | 3–5.99 | 6–8.99 | 9–12.99 | >13 | No data

Note: The District of Columbia boasts a higher rate than any other region, with 15.9 homicides per 100,000 people.

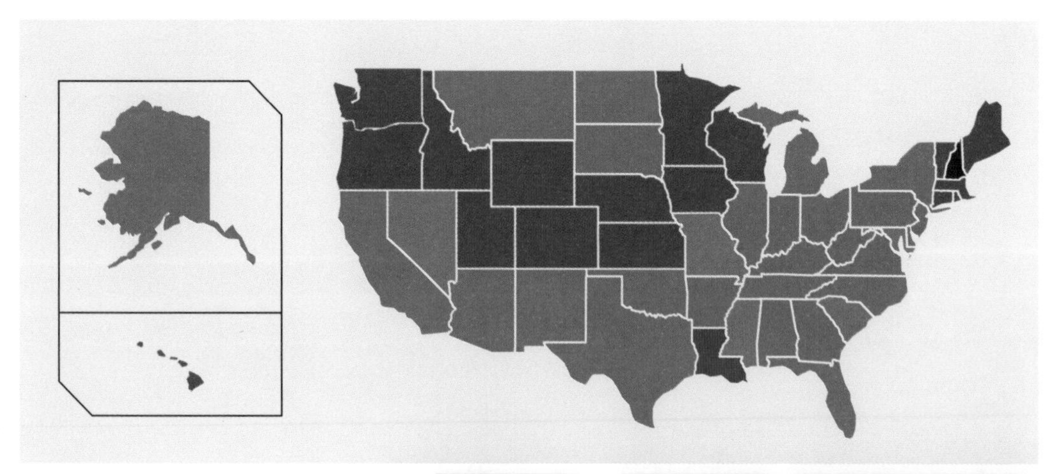

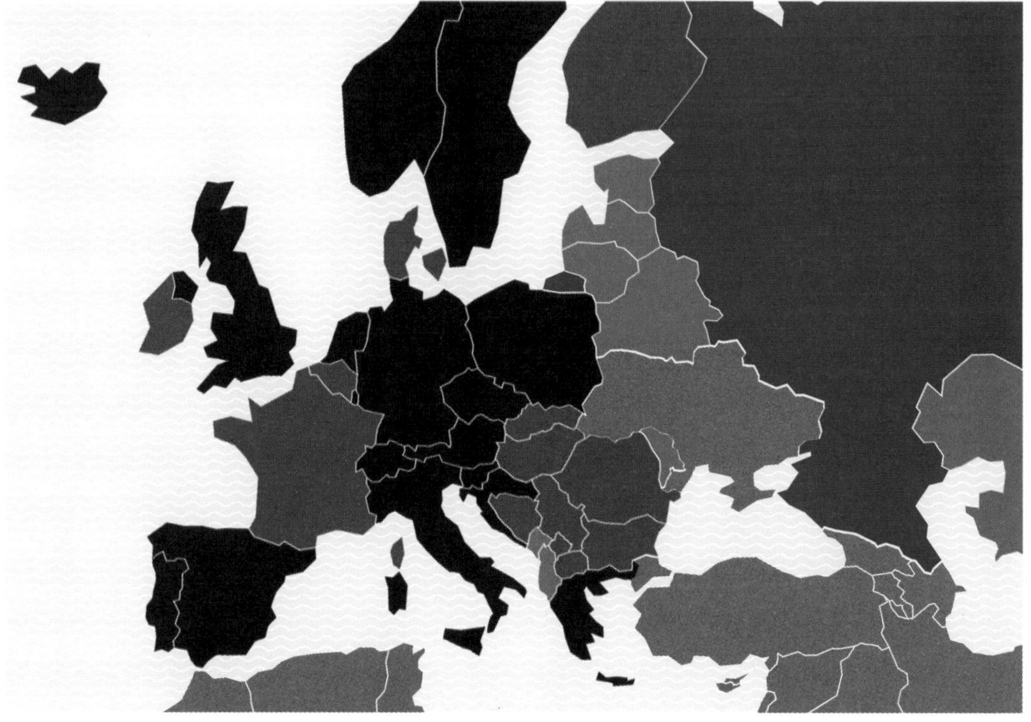

* Including Russia

90 Number of **executions** since 1976 in the US*

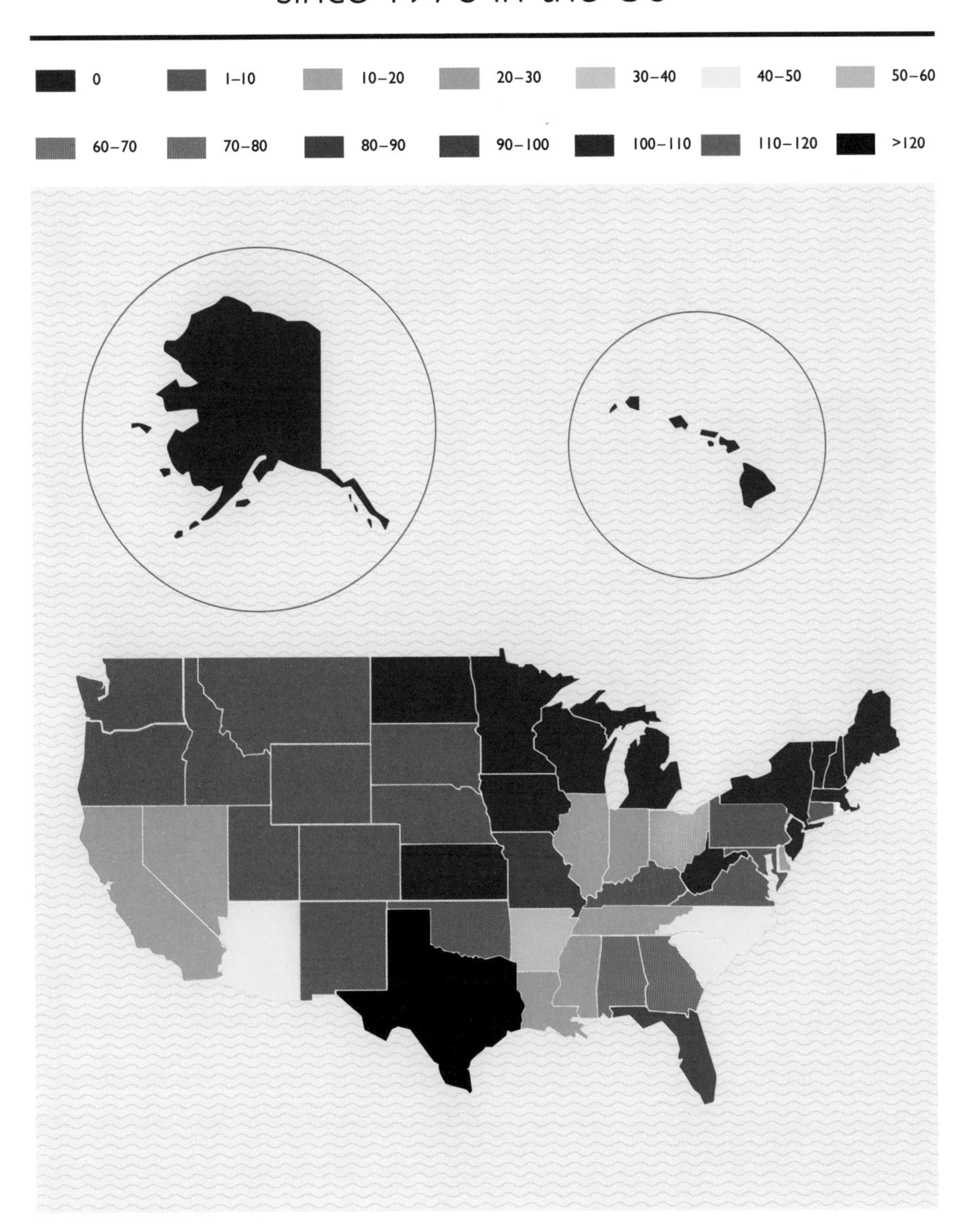

* Accurate as of February 16, 2018

91 Capital punishment
laws of the world

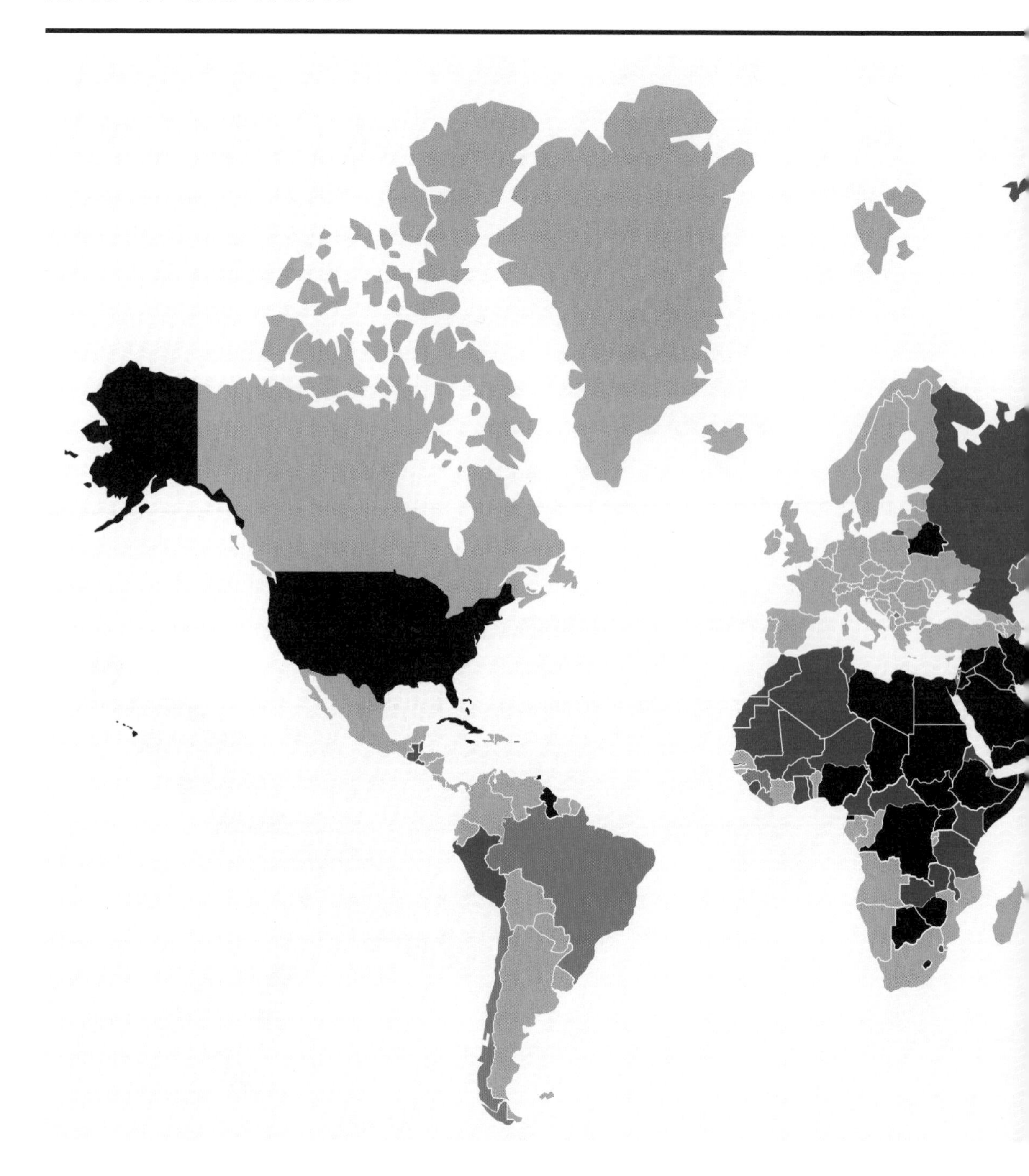

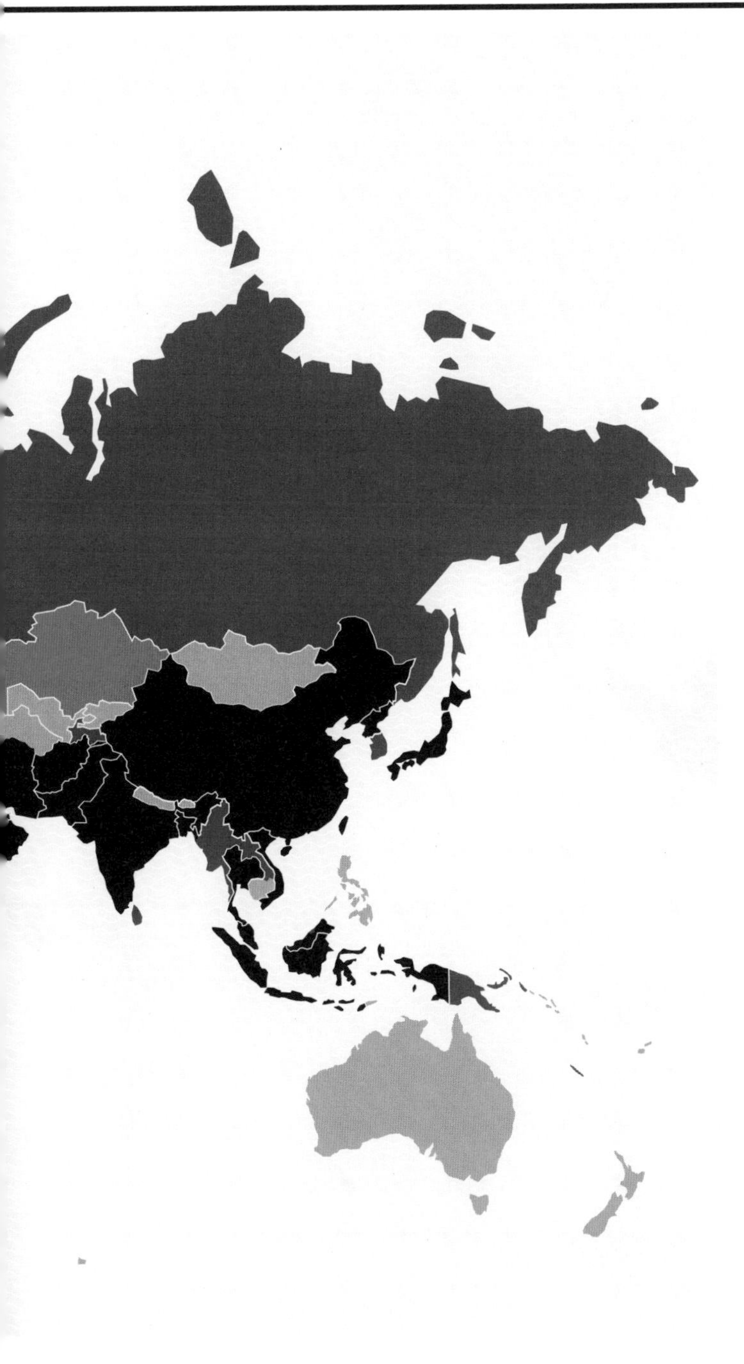

Abolished for all crimes

Abolished for ordinary crimes; death penalty used in exceptional circumstances

Abolished in practice; retains the death penalty but has not executed anyone in the last ten years

Retained

92 Prison population
per 100K people

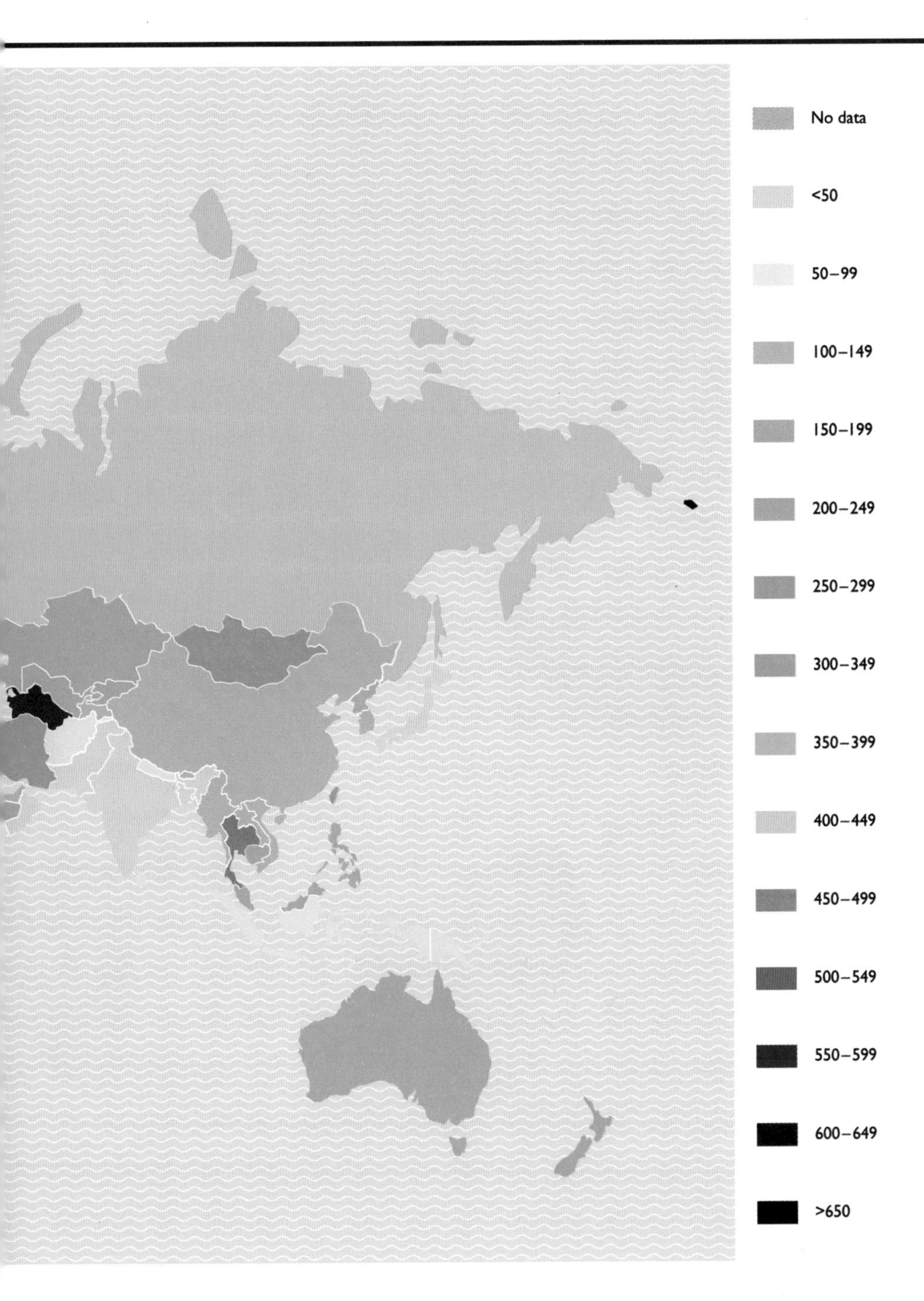
No data
<50
50–99
100–149
150–199
200–249
250–299
300–349
350–399
400–449
450–499
500–549
550–599
600–649
>650

93 Every recorded **terrorist attack** between 1970 and 2015*

* With or without deaths; carried out or not

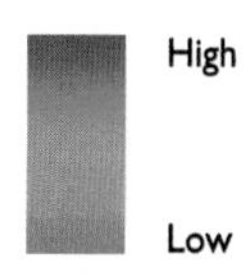

Intensity value is a combination of incident fatalities and injuries.

9
NATURE

94 Countries with **no rivers**

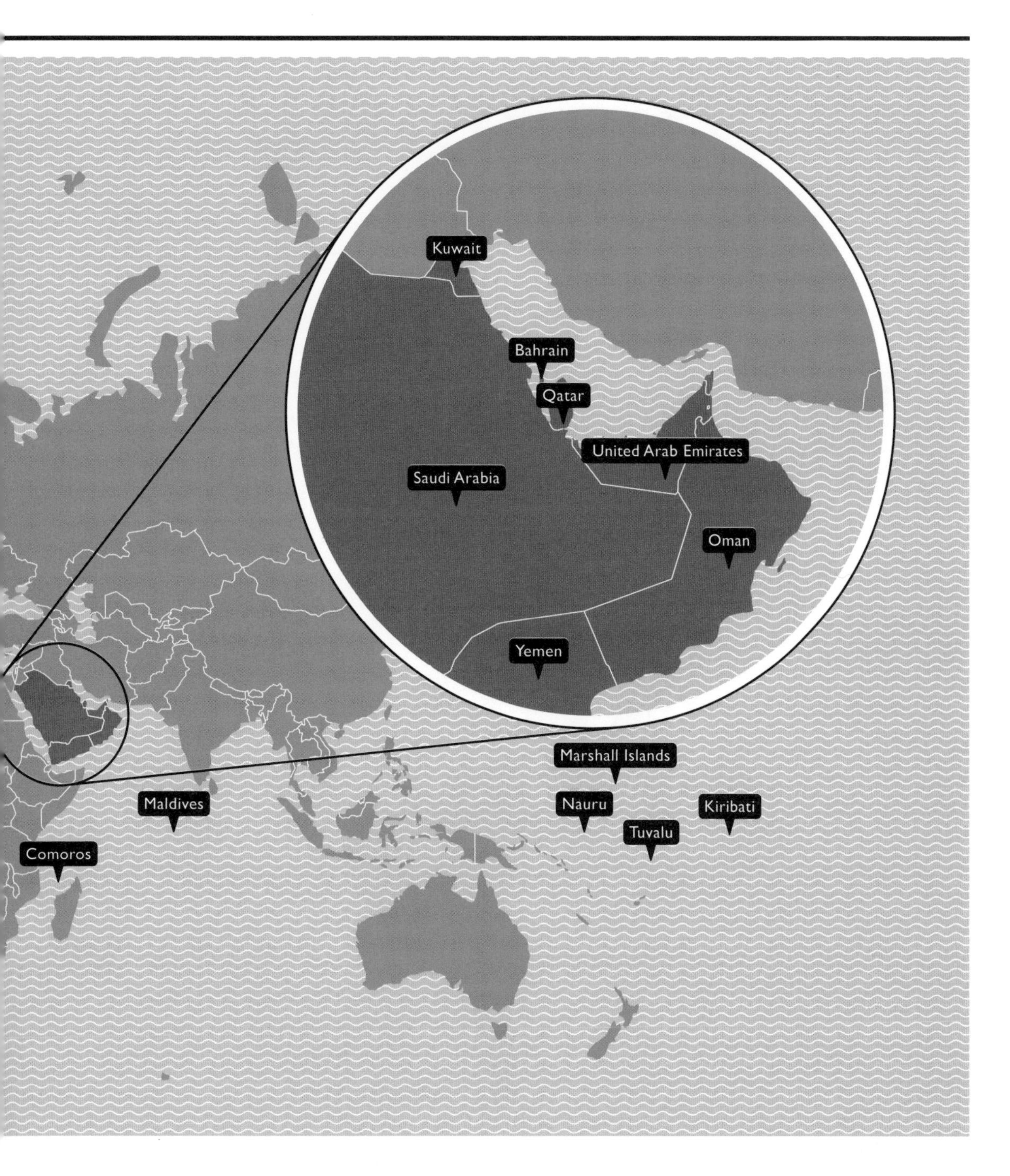
Kuwait
Bahrain
Qatar
United Arab Emirates
Saudi Arabia
Oman
Yemen
Marshall Islands
Maldives
Nauru
Kiribati
Tuvalu
Comoros

95 Countries with the most **venomous animals**

Number of venomous animal species

- <10
- 10–20
- 20–30
- 30–40
- 40–50
- >50
- No data

Which country has the most venomous animals? Surprisingly, it's not Australia but Mexico with eighty, followed by Brazil with seventy-nine, and then Australia with only sixty-six (they just happen to be more potent). If you're wondering about France's high rating, it's because France includes French Guiana, which is considered an overseas department of France.

96 Historic vs. present geographical distribution of **lions** (*Panthera leo*)

Historic distribution

Present distribution

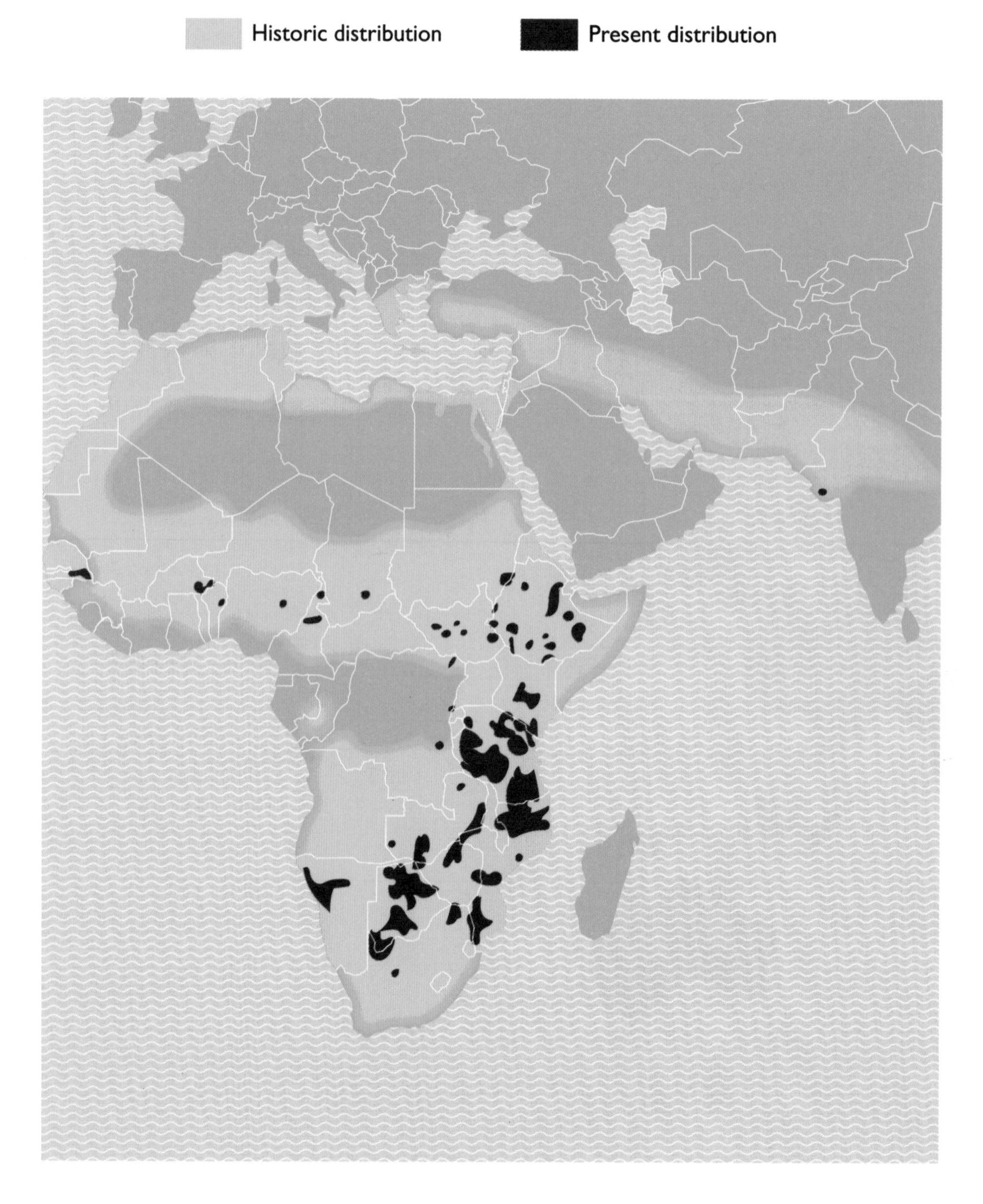

97 How many Americans believe that **climate change** will affect them?

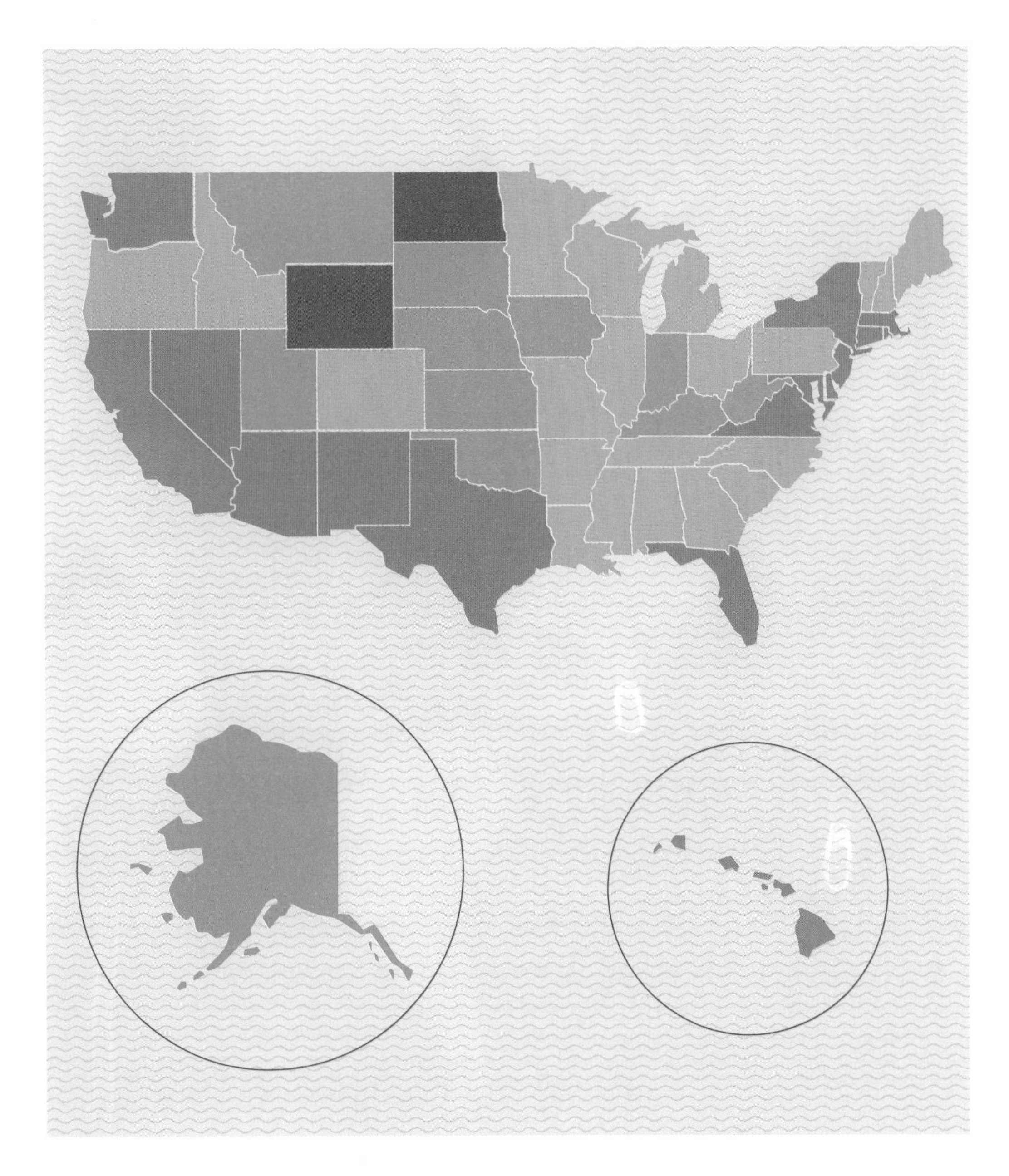

98 All the **sharks killed by humans** vs. all the **humans killed by sharks** in 2017

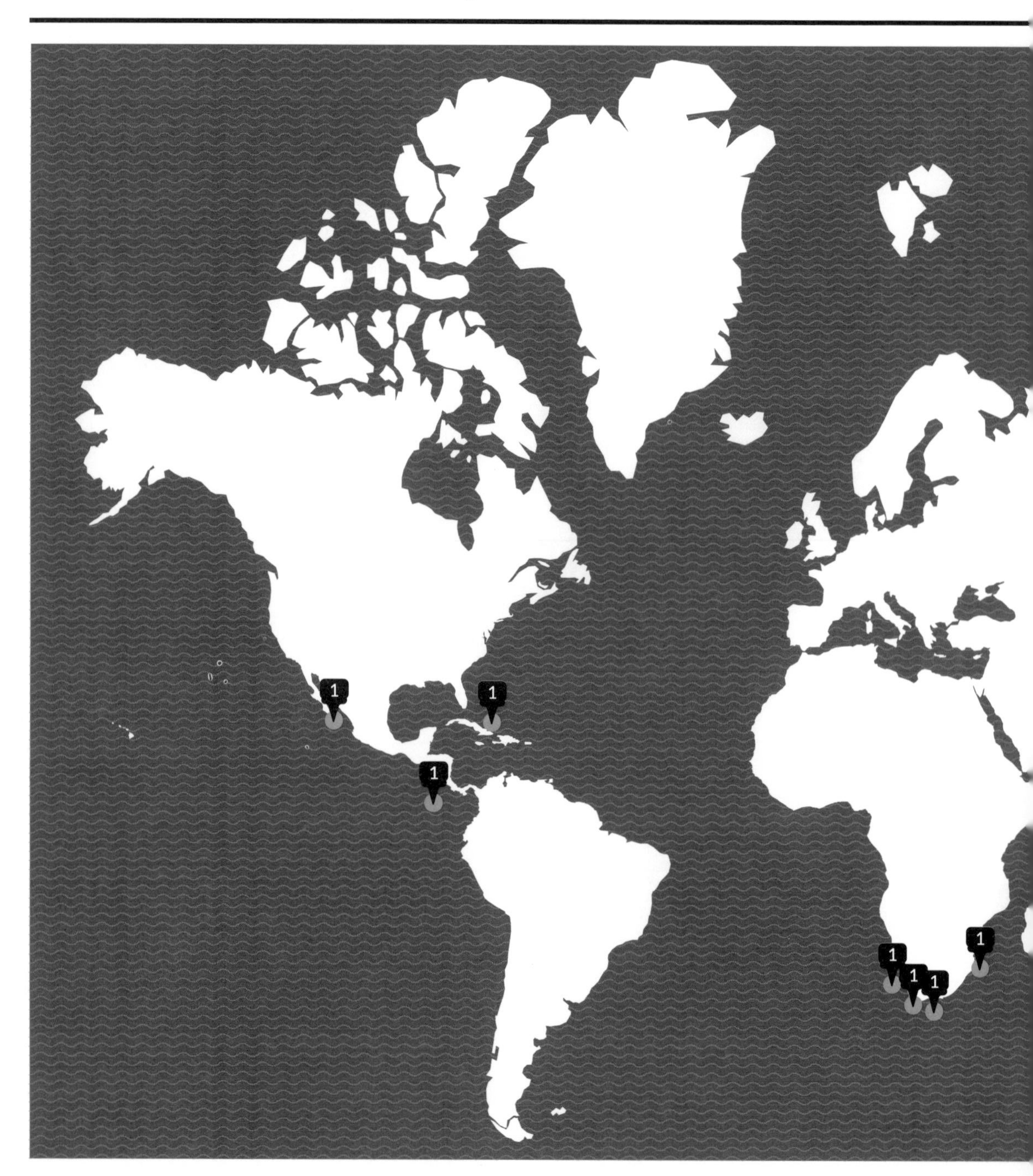

Sharks killed by humans:
63,000,000–274,000,000

Humans killed by sharks:
7–10

Global Shark Attack File notes that in 2017 a further thirty-one migrants crossing the Mediterranean Sea may have been killed by sharks, though this is disputed by experts.

99 Probability of a **white Christmas**

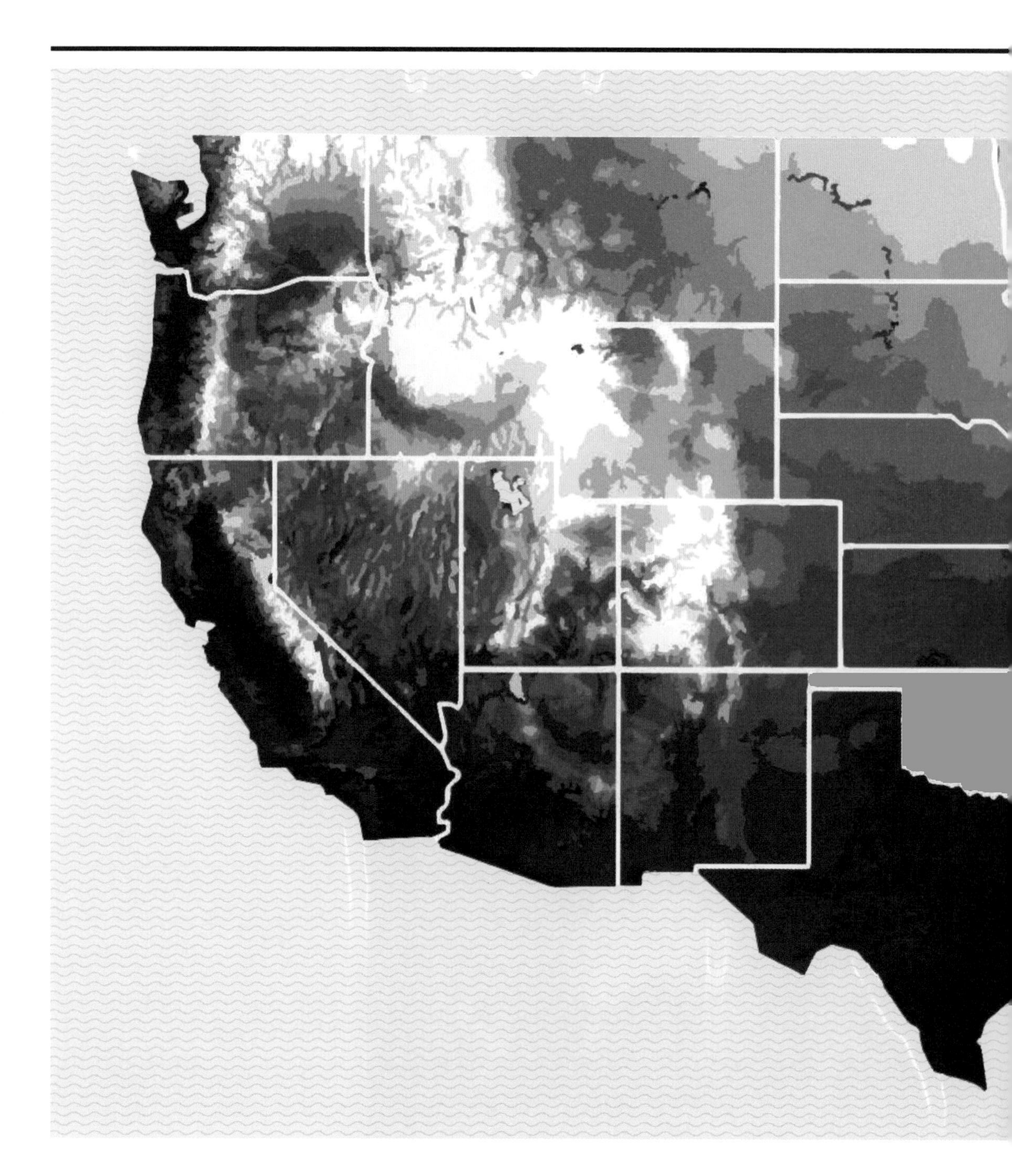

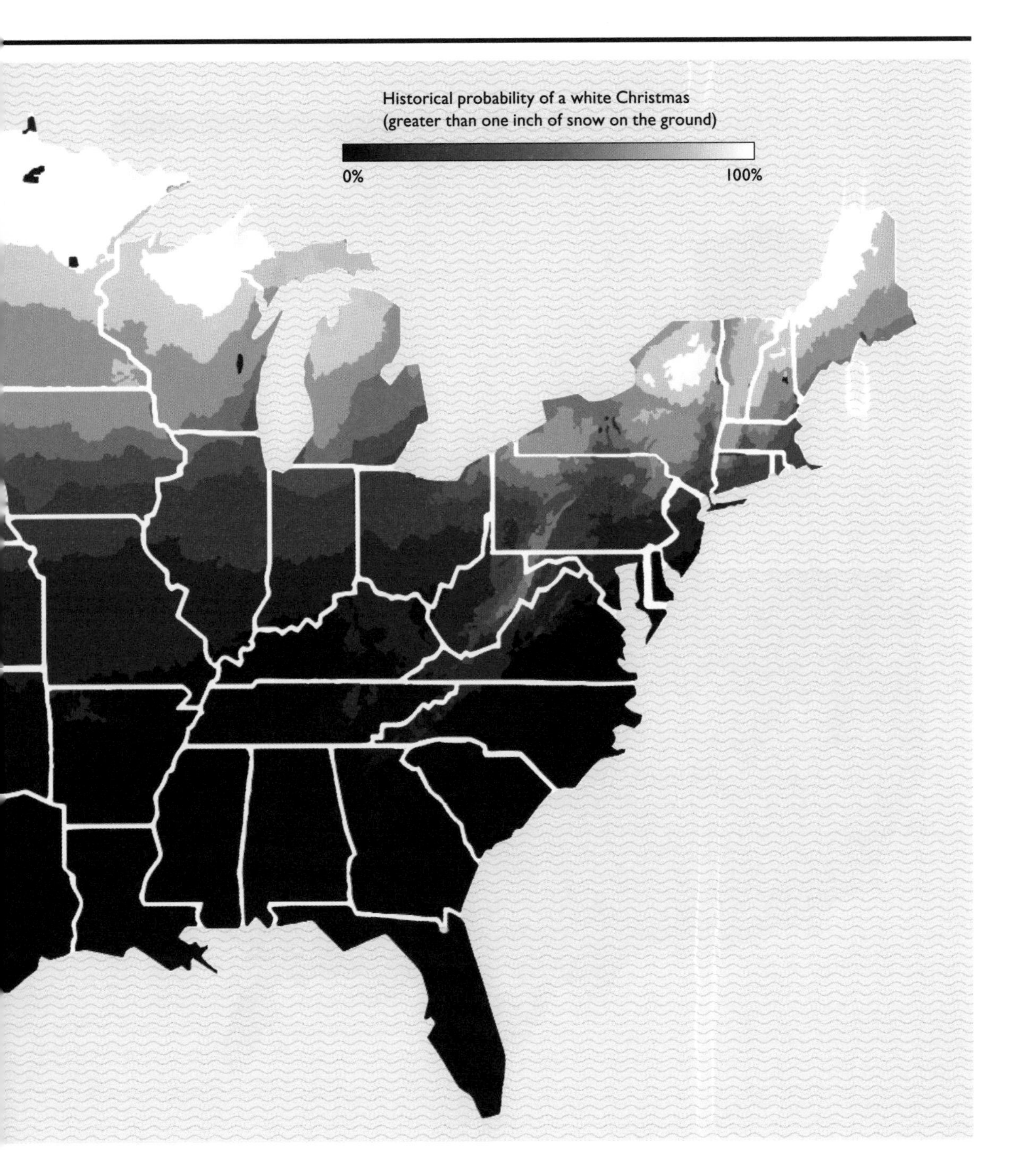
Historical probability of a white Christmas
(greater than one inch of snow on the ground)
0%
100%

100 Average annual hours of **sunlight** in the US and in Europe

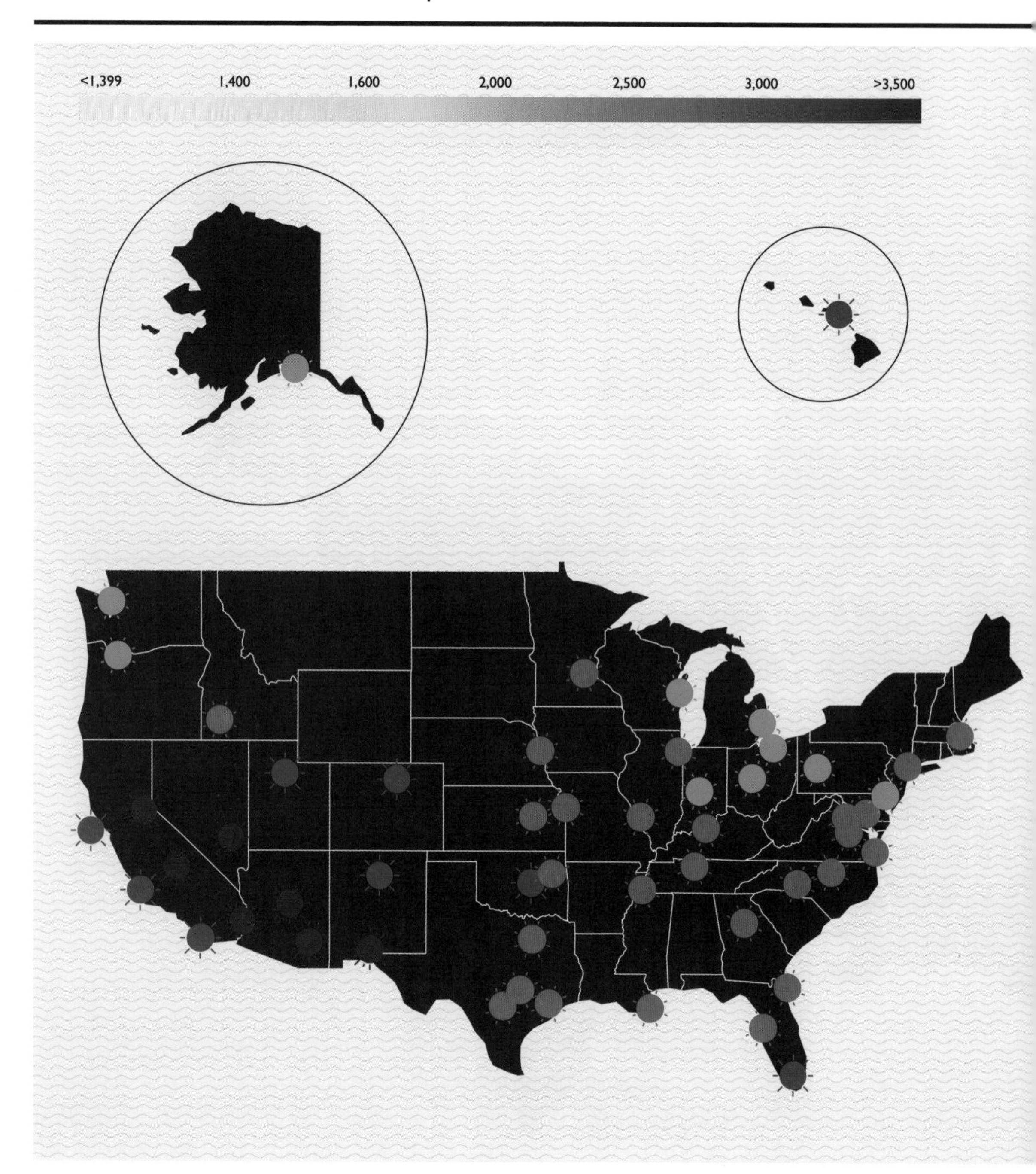

SOURCES

People and Populations

1. European countries overlaid on areas of the Americas with equal populations
Data: *The World Factbook*, Central Intelligence Agency. Used with the permission of the CIA.
Concept: Redditor Speech500.

2. US states overlaid on areas of Europe with equal populations
Data: *The World Factbook*, Central Intelligence Agency. Used with the permission of the CIA.
Concept: Redditor Speech500.

3. How the North American population fits into Europe
Data: Copyright © 2017 World Population Review, worldpopulationreview.com.
Concept: Redditor jackblack2323.

4. The astounding drop in global fertility rates from 1970 to 2015
Data: United Nations, Department of Economic and Social Affairs, Population Division (2017), *World Population Prospects: The 2017 Revision,* DVD Edition.
Concept: Redditor Areat. He is from France and his work on maps has highlighted changing global perspectives.

5. More than half of the Australian population lives here
Data: Copyright © Commonwealth of Australia. Licensed under a Creative Commons Attribution 2.5 Australia license.
Concept: Redditor Ianson15.

6. 50% of Canadians live south of the red line
Data: All rights reserved © 2018 ForestWalk, Inc.
Concept: Robin Lyster.

7. Countries with the largest immigrant populations
Data: United Nations, Department of Economic and Social Affairs (2017), "International Migrant Stock: The 2017 Revision." Used with the permission of the United Nations.
Concept: Wikipedia contributor Stephen Bain, "Countries by Immigrant Population," Wikimedia Commons 2018 under Creative Commons license CC BY-SA 3.0.

8. Second-largest nationality living in each European country
Data: United Nations, Department of Economic and Social Affairs (2017), "International Migrant Stock: The 2017 Revision." Used with the permission of the United Nations.
Concept: Redditor JimWillFixIt69.

9. Percentage of people born in each European country now living abroad
Data: United Nations, Department of Economic and Social Affairs (2017), "International Migrant Stock: The 2017 Revision." Used with the permission of the United Nations.
Concept: Max Galka, an information designer, adjunct lecturer at the University of Pennsylvania's School of Design, contributor to *Guardian's* Cities, founder of FOIA Mapper, and an advocate for crypto market transparency via Elementus.

10. Countries and overseas territories that each have smaller populations than Greater Tokyo
Data: *The World Factbook*, Central Intelligence Agency. Used with the permission of the CIA.
Concept: Redditor, xtehshadowx.

11. World median ages
Data: *The World Factbook*, Central Intelligence Agency. Used with the permission of the CIA.
Concept: Simran Khosla. She comes from a visual journalism background and loves experimenting with new forms of storytelling online. As a journalist, she has worked on infographics, data visualization, and animated explainers. As a developer, she hopes to keep building beautiful things with a strong sense of narrative on the web.

12. Average female height worldwide
Data: Copyright © Wikipedia 2018, "Average Human Height Worldwide," under Creative Commons Attribution-ShareAlike 3.0 Unported License.

13. Average male height worldwide
Data: Copyright © Wikipedia 2018, "Average Human Height Worldwide," under Creative Commons Attribution-ShareAlike 3.0 Unported License.

Politics, Power, and Religion

14. 64 countries have had a female leader in the last 50 years
Data and concept: "Number of Women Leaders Around the World Has Grown, but They're Still a Small Group," Pew Research Center, Washington, DC (March 8, 2017), pewresearch.org/fact-tank/2017/03/08/women-leaders-around-the-world

15. Countries with state religions
Data: Copyright © Wikipedia 2018, "State religion," under Creative Commons Attribution-ShareAlike 3.0 Unported License
Concept: Wikimedia contributor Smurfy, Wikimedia Commons under Creative Commons license CC0 1.0

16. Fastest-growing religion in each country around the world
Data: "The Future of World Religions: Population Growth Projections, 2010–2050." Pew Research Center, Washington, D.C. (April 10, 2015) pewforum.org/2015/04/02/religious-projections-2010-2050.

17. Simplified map of Africa's religions
Data: "Tolerance and Tension: Islam and Christianity in Sub-Saharan Africa," Pew Research Center, Washington, DC (April 10, 2015), pewforum.org/2010/04/15/executive-summary-islam-and-christianity-in-sub-saharan-africa.

18. England vs. Great Britain vs. United Kingdom
Concept: TW Carlson, "Euler Diagram of the British Isles.svg," Wikimedia Commons 2018 under CCO 1.0 Universal Public Domain Dedication.

19. Birthplaces of religious leaders
Original concept

20. Every country's highest-valued export
Data: *The World Factbook*, Central Intelligence Agency. Used with the permission of the CIA.
Concept: Simran Khosla.

21. The largest source of imports by country
Data: Alexander James Gaspar Simoes and César A. Hidalgo, "The Economic Complexity Observatory: An Analytical Tool for Understanding the Dynamics of Economic Development," paper from the 2011 AAAI Workshop on Scalable Integration of Analytics and Visualization.
Concept: Redditor Amiantedelxue, a French geography student and map aficionado.

22. Countries with economies bigger than California's
Data: "California's Economy Is Now the 5th-Biggest in the World, and Has Overtaken the United Kingdom," *Business Insider* (May 5, 2018), businessinsider.com/california-economy-ranks-5th-in-the-world-beating-the-uk-2018-5?r=US&IR=T.

23. World gold reserves in grams per person
Data: Copyright © 2019 World Gold Council, gold.org.
Concept: www.merchantmachine.co.uk.

24. World divided in half based on military spending
Data: Copyright © SIPRI 2018, sipri.org/databases/milex.
Concept: Philip Kearney, an amateur cartographer based in Austin, Texas. Before he could tie his shoes, he was creating maps of his neighborhood to share with friends. These days, Philip creates unique and interesting maps to share online, while also completing degrees in programming and computer-based mapping. philip-kearney.com.

25. Nuclear vs. non-nuclear countries
Data: Copyright © 2018 International Atomic Energy Agency (IAEA). All rights reserved.
Concept: britishbusinessenergy.co.uk.

Culture and Customs

26. Who are the world's speed demons? The highest speed limits around the world
Data: Copyright © Wikipedia 2018, "Speed Limits by Country," under Creative Commons Attribution-ShareAlike 3.0 Unported License.
Concept: Wikimedia contributor Amateria1121, "World Speed Limits.svg," 2014, under Creative Commons Attribution-Share Alike 4.0 International license.

27. Who drives on the "wrong" side of the road?
Data: Copyright © Wikipedia 2018, "Left- and Right-Hand Traffic," under Creative Commons Attribution-ShareAlike 3.0 Unported License.
Concept: Benjamin D. Esham, 2007 under public domain.

28. Football vs. soccer
Data: Copyright © Wikipedia 2018, "Names for Association Football," under Creative Commons Attribution-ShareAlike 3.0 Unported License.
Concept: Redditor reddripper. Arif, a programmer and graphic designer from Indonesia, works with databases and web programming, as well as with cover design and illustration.

29. Most commonly spoken languages after English and Spanish
Data: US Census Bureau's American Community Survey.
Concept: slate.com.

30. Map of countries officially not using the metric system
Data: *The World Factbook*, Central Intelligence Agency.
Concept: Wikimedia contributor AzaToth, updated by Cherkash, under public domain.

31. Decimal point vs. decimal comma vs. other decimal separators
Data: Copyright © Wikipedia 2018, "Decimal Separator," under Creative Commons Attribution-ShareAlike 3.0 Unported License.
Concept: Wikimedia contributor NuclearVacuum, "DecimalSeparator.svg," licensed under CC BY-SA 3.0.

32. How to write the date in different countries
Data: Copyright © Wikipedia 2018, "Date Format by Country," under Creative Commons Attribution-ShareAlike 3.0 Unported License.
Concept: Wikimedia contributor Typhoon2013, "Date Format by Country (New).png," Licensed under CC BY-SA 4.0.

33. Election days by country
Data: Copyright © Wikipedia 2018, "Election Day," under Creative Commons Attribution-ShareAlike 3.0 Unported License.
Concept: Renno Hokwerda, a student of planning and geography from the Netherlands. His maps feature curious, thrilling sidenotes in geography, and he produces highly detailed maps of (realistic) imaginary cities, which can be found online. flickr.com/photos/31322479@N04/albums/72157624717280387.

34. Countries that have no McDonald's
Data: Copyright © Wikipedia 2018, "List of Countries with McDonald's Restaurants," under Creative Commons Attribution-ShareAlike 3.0 Unported License.
Concept: Wikimedia contributor Szyslak, 2007.

35. Cats vs. dogs
Data: Google Keyword Planner, September 2015.
Original concept

36. The most photographed places in the world
Data and concept: Google Maps's Panoramio service, a now-defunct aggregator of user-uploaded geolocated images.

37. Heavy metal bands per 100K people
Data: Copyright © 2002–2019 Encyclopaedia Metallum. *The World Factbook*, Central Intelligence Agency. Used with the permission of the CIA.
Concept: Redditor depo_.

38. Countries with the most Miss World winners
Original concept

39. Longest place names
Data: Copyright © Wikipedia 2018, "Longest Place Names," under Creative Commons Attribution-ShareAlike 3.0 Unported License.
Concept: Jay Bhadrica.

40. Most recurring word on each country's English Wikipedia page
Data and concept: Redditor Amiantedelxue.

41. Age of consent for heterosexual sex
Data and concept: Copyright © Wikimedia contributor MissMJ, Wikimedia Commons, under CC0 1.0 Universal (CC0 1.0) Public Domain Dedication.

42. Male circumcision: one thing that unites the US and the Middle East
Data and concept: Copyright © Morris, Wamai, Henebeng, Tobian, Klausner, Banerjee & Hankins, 2016 CC BY 4.0.

43. World plug and socket map
Data: Copyright © 2003–2019 World Standards, worldstandards.eu/electricity/spread-plug-types-map.
Concept: Conrad H. McGregor, who started the website worldstandards.eu in 2003 as a source of information for travelers or anyone who is interested in the subject of international standards.

Friends and Enemies

44. Open borders of the world
Data: Copyright © Wikipedia 2018, "Open Borders," under Creative Commons Attribution-ShareAlike 3.0 Unported License.
Concept: Redditor Fweepi.

45. Who Americans consider their allies, friends, and enemies
Data: Copyright © yougov.com, 2017.
Concept: Redditor ShilohShay.

46. Countries the US is obligated to go to war for (for now)
Data: US Department of State 2018, US Collective Defense Arrangements, state.gov/s/l/treaty/collectivedefense (public domain).

47. 22 countries that the UK has not invaded
Data: Stuart Laycock (2012), *All the Countries We've Ever Invaded: And the Few We Never Got Round To*, History Press.

48. Countries that were raided or settled by the Vikings
Data: Copyright © Wikipedia 2018, "Viking Expansion," under Creative Commons Attribution-ShareAlike 3.0 Unported License.
Concept: Redditor Grankogle, from Denmark, with an interest in history, maps, culture, and other curiosities.

49. European countries that have invaded Poland
Data: Copyright © Wikipedia 2018, "List of Wars Involving Poland," under Creative Commons Attribution-ShareAlike 3.0 Unported License.
Concept: Redditor ClayTownR.

50. Zone Rouge: an area of France so badly damaged during World War I that people were forbidden to live there
Data: J. Guicherd and C. Matriot (1921), "La terre des régions dévastées," *Journal d'Agriculture Pratique* 34, 154–56.
Concept: Tinodela and Lamiot, Wikimedia Commons under Creative Commons license CC BY-SA 4.0.

51. Countries that officially recognize the State of Palestine
Data: Copyright © Wikipedia 2018, "International Recognition of the State of Palestine," under Creative Commons Attribution-ShareAlike 3.0 Unported License.
Concept: Wikimedia contributor Amateria1121, Wikimedia Commons, 2012 under CC BY-SA 4.0.

52. Countries that officially recognize the State of Israel
Data: Copyright © Wikipedia 2018, "International Recognition of Israel," under Creative Commons Attribution-ShareAlike 3.0 Unported License.
Concept: Wikimedia contributor AMK1211, 2011, under Wikimedia Commons license CC BY-SA 4.0.

53. Where North Korea has embassies
Data: Copyright © Wikipedia 2018, "List of Diplomatic Missions of North Korea," under Creative Commons Attribution-ShareAlike 3.0 Unported License.
Concept: Wikimedia contributor Avala, 2008, under public domain.

54. Who has embassies in North Korea?
Data: Copyright © Wikipedia 2018, "List of Diplomatic Missions of North Korea," under Creative Commons Attribution-ShareAlike 3.0 Unported License.
Concept: Copyright © Kwamikagami, Wikimedia Commons under Creative Commons license CC BY-SA 4.0, copyright © Citynoise Wikimedia Commons under Creative Commons license CC BY-SA 4.0.

55. Former and current communist countries
Original concept

Geography

56. Mercator projection vs. the true size of countries
Concept: Redditor and climate data scientist neilrkaye.

57. Chile is a ridiculously long country
Data: thetruesize.com.

58. The true size of Africa
Data: thetruesize.com.
Concept: Redditor edtheredted.

59. The Pacific Ocean is larger than all the land on Earth
Data: Made with Natural Earth. Free vector and raster map data from naturalearthdata.com.
Concept: Chris Stephens, from naturalearthdata.com.

60. The 20 largest islands in the world compared
Data: Copyright © Wikipedia, 2018, "Lists of Islands by Area," under Creative Commons Attribution-ShareAlike 3.0 Unported License.

61. The Pan-American Highway: the longest road in the world
Data: Copyright © NASA, 2015.
Concept: Wikimedia contributor Seaweege, Wikimedia Commons, under CC0 1.0 Universal (CC0 1.0) Public Domain Dedication.

62. More people live inside this circle than outside of it
Data: *The World Factbook*, Central Intelligence Agency. Used with the permission of the CIA.
Concept: Redditor valeriepieri.

63. Antipodes world map—or, why you can't dig to China from the US
Data: Copyright © Kwamikagami, Wikimedia Commons, under Creative Commons license CC BY-SA 4.0, copyright © Citynoise Wikimedia Commons, under Creative Commons license CC BY-SA 4.0.
Concept: Redditor mattsdfgh.

64. The world in a mirror
Concept: Redditor Ambamja.

65. The world's time zones
Data: Copyright © Wikipedia 2018, "List of Time Zones by Country," under Creative Commons Attribution-ShareAlike 3.0 Unported License.
Concept: Branden Rishel lives with his wonderful family in Bellingham, Washington: cartographerswithoutborders.org.

66. The world's five longest domestic nonstop flights
Data: Google maps.
Concept: Redditor bonne-nouvelle.

67. Travel times from London in 1914
Data: Copyright © Rome2rio, 2018.
Concept: Rome2rio, a door-to-door travel information and booking engine, helping people get to and from any location in the world.

68. Travel times from London in 2016
Data: Copyright © Rome2rio, 2018.
Concept: Rome2rio, a door-to-door travel information and booking engine, helping people get to and from any location in the world.

69. Luxembourg is not a microstate!
Data: thetruesize.com.
Concept: Redditor issoweilsosoll.

70. All roads lead to Rome
Concept: roadstorome.moovellab.com.

Benedikt Groß is an antidisciplinary designer working at the intersection of people, their data, technology and environments, somewhere in the Bermuda Triangle of data, speculative, and computational design. He is also professor of interaction design at HfG Schwäbisch Gmünd and director of design at moovel lab.

Raphael Reimann is a multidisciplinary urbanist and founding member of moovel lab. His background is in geography and urban design and development. He works on ideation, conceptualization, and communication of moovel lab's projects. His work, at the intersection of fast-paced digital services and persistent city infrastructure, is consistently interesting and challenging.

Philipp Schmitt is a designer and artist interested in relationships of technology and society. His work revolves around design and technology as subjects instead of considering them solely as tools. He works collaboratively across disciplines, drawing from interaction and generative design, speculative design, data visualization, photography, and filmmaking, among others.

History

71. European map of the unexplored world (1881)
Data: Jules Verne (1881), *The Great Explorers of the Nineteenth Century*, archive.org/details/greatexplorersof00vernuoft.

72. Colonial Africa on the eve of World War I
Data: Copyright © Eric Gaba, Wikimedia Commons 2018, "Colonial Africa 1913, Pre WWI.svg," licensed under CC BY-SA 4.0.
Concept: Minas Giannekas, the developer behind mapchart.net, the most popular map-creating website.

73. Map of the entire internet in December 1969
Data: ARPAnet geographical maps (1969–1986), taken from the Computer History Museum archives, computerhistory.org/collections/catalog/102646702.

74. If the Roman Empire reunited—using modern borders
Data: Copyright © Wikipedia 2018, "Roman Empire," under Creative Commons Attribution-ShareAlike 3.0 Unported License.
Concept: Redditor Fweepi.

75. The first proposed map of Pakistan and the partition of India
Map in general use.

76. If the Mongol Empire reunited
Data: Copyright © Wikipedia 2018, "Mongol Empire," under Creative Commons Attribution-ShareAlike 3.0 Unported License.
Concept: Redditor Trapper777_.

77. When Great Britain was connected to continental Europe
Data: "Doggerland—The Europe That Was," © 1996–2019 National Geographic Society. All rights reserved. Used with permission from National Geographic.

78. Where were the Seven Wonders of the Ancient World?
Original concept

79. WWI casualties as a percentage of prewar populations
Data: Copyright © Wikipedia 2018, "World War I Casualties," under Creative Commons Attribution-ShareAlike 3.0 Unported License.
Concept: Redditor Ianson15.

80. WWII casualties as a percentage of prewar populations
Data: Copyright © Wikipedia 2018, "World War II Casualties," under Creative Commons Attribution-ShareAlike 3.0 Unported License.

81. Countries that lost citizens on 9/11
Data: Copyright © Wikipedia, 2018, "Casualties of the September 11 Attacks," under Creative Commons Attribution-ShareAlike 3.0 Unported License.
Concept: Redditor thepenaltytick.

National Identity

82. Colors of passports around the world
Data: Copyright © Wikipedia 2018, "List of Passports," under Creative Commons Attribution-ShareAlike 3.0 Unported License.
Concept: Wikipedia contributor Twofortnights, Wikimedia Commons under Creatve Commons license CC BY-SA 4.0.

83. Countries whose flags contain red and/or blue
Data: *The World Factbook*, Central Intelligence Agency. Used with the permission of the CIA.
Concept: Redditor PieJesu.

84. Flags of the world
Data: Copyright © FOTW, Flags of the World, flagspot.net/flags.
Concept: Douglas Wilhelm Harder, a contributing lecturer at the University of Waterloo in the Department of Electrical and Computer Engineering since 2002. He completed his undergraduate degree at University of Guelph. He was a master corporal in the Canadian Forces Reserve with the Lincoln and Welland Regiment and 2 Intelligence Company. He previously worked as a mathematical software developer at Maplesoft Inc. following the completion of his masters in applied mathematics at the University of Waterloo.

85. If European borders were drawn by DNA instead of ethnicity
Data: Copyright © 2004–2018 Eupedia.com. All Rights Reserved.

86. "Indian" isn't a language
Original concept

87. What do we mean when we say "Asian"?
Original concept

Crime and Punishment

88. The US has as many murders annually as all the countries in blue combined
Data: United Nations Office on Drugs and Crime, "Global Study on Homicide—Statistics and Data," dataunodc.un.org.
Concept: Redditor Rift3N.

89. Homicide rates: Europe vs. the US
Data: Copyright © ucrdatatool.gov, an official site of the US Federal Government, US Department of Justice.

90. Number of executions since 1976 in the US
Data: Copyright © 2018 Death Penalty Information Center.
Concept: Redditor lursh123.

91. Capital punishment laws of the world
Data and concept: "Death Penalty" © 2018 Amnesty International, amnesty.org/en/what-we-do/death-penalty.

92. Prison population per 100K people
Data: Copyright © World Prison Brief, Institute for Criminal Policy Research.
Concept: Redditor Spartharios.

93. Every recorded terrorist attack between 1970 and 2015
Data: Global Terrorism Database developed by the National Consortium for the Study of Terrorism and Responses to Terrorism (START) at the University of Maryland. Copyright University of Maryland 2018.
Concept: START is a university-based research, education, and professional training center comprising an international network of scholars committed to the scientific study of the causes and human consequences of terrorism in the United States and around the world, as well as societal responses to terrorism. start.umd.edu and start.umd.edu/gtd.

Nature

94. Countries with no rivers
Data: Copyright © Wikipedia 2018, "List of Countries Without Rivers," under Creative Commons Attribution-ShareAlike 3.0 Unported License.
Concept: Redditor darth_stroyer.

95. Countries with the most venomous animals
Data: Copyright © Armed Forces Pest Management Board, 2018.
Concept: Redditor lanson15.

96. Historic vs. present geographical distribution of lions (*Panthera leo*)
Data and concept: Copyright © Lion ALERT 2018. Data sourced from African Lion & Environmental Research Trust: a responsible development approach to lion conservation. lionalert.org.

97. How many Americans believe that climate change will affect them?
Data: Peter D. Howe, Matto Mildenberger, Jennifer R. Marlon, and Anthony Leiserowitz (2015). "Geographic Variation in Opinions on Climate Change at State and Local Scales in the USA," *Nature Climate Change*, 5.
Concept: The Yale Climate Opinion Maps were created by J. Marlon and A. Leiserowitz (Yale Program on Climate Change Communication), P. Howe (Utah State University), and M. Mildenberger (University of California, Santa Barbara).

98. All the sharks killed by humans vs. all the humans killed by sharks in 2017
Data: Copyright © 2018 Elsevier B.V.
Concept: Branden Rishel.

99. Probability of a white Christmas
Data: Licensed from weatherspark.com. Copyright © Cedar Lake Ventures, Inc. All rights reserved.
Concept: Janne Peuhkuri.

100. Average annual hours of sunlight in the US and in Europe
Data: Copyright © Wikipedia 2018, "List of Cities by Sunshine Duration," under Creative Commons Attribution-ShareAlike 3.0 Unported License.
Concept: jesusgonzalezfonseca.blogspot.com.

ACKNOWLEDGMENTS

Starting a map-based website or blog was relatively easy. What's more difficult is keeping it going. And what's been a real challenge is translating a website into a book. On June 21, 2016 (just two days before the Brexit Referendum) I received an email from Laura Barber at Granta Books asking if I had any plans to bring out a book. I'd been approached by publishers before, but whenever I'd explained that I didn't hold all the rights to the designs of the maps, they had always backed out. Laura, to her everlasting credit, came up with an ingenious solution. Instead of publishing the most popular maps from the website as they appeared, we would reimagine and redraw the maps with a consistent style and updated data. Granta would contact all of the original cartographers—whose biographies you can read at the back of this book—to ensure everyone was happy. Granta would also include a few entirely new maps, so that even diehard fans of the website would have something fresh.

Creating over a hundred maps from scratch is a major task, and since I'm not a graphic designer, I was not the one to do it. That's why we've used the team from Infographic.ly. They've created some stunning maps with a stylish, distinctive look. Credit for overseeing the mapmaking process goes to Ka Bradley, Sinéad O'Callaghan, and Jay Bhadricha at Granta, who also gave a huge amount of helpful feedback about each of the designs. Again, this book could not have been published without them. Ka in particular deserves a huge amount of credit since she not only helped to manage the design process but also managed me. And for the North American edition, I would like to thank the team at the Experiment.

Just one final word. We have done our best to fix any mistakes that appeared on the original maps, but if the feedback on my website is anything to go by, it is possible there might be a few errors. While putting this book together has been a collaboration, I've had the final say over each map, so any errors are mine and mine alone.

ABOUT THE AUTHOR

IAN WRIGHT runs Brilliant Maps, one of the most popular cartographic sites on the internet. In addition to being a cartophile, he's also a keen walker. In 2015, he combined these two passions to become the first person to walk all of the newly expanded London Tube map. Originally from Canada, he now lives in the UK.

BRILLIANTMAPS.com

INFOGRAPHIC.LY is an infographic and information design agency based in Dubai.

WASHINGTON
MONTANA
NORTH DAKOTA
OREGON
IDAHO
WYOMING
SOUTH DAKOTA
NEBRASKA
NEVADA
UTAH
COLORADO
KANSAS
CALIFORNIA
ARIZONA
NEW MEXICO
OKLAHOMA
TEXAS
ALASKA
HAWAII